19.35

Sociobiology and Behavior

Sociobiology and Behavior

DAVID P. BARASH

Departments of Psychology and Zoology
University of Washington

Foreword by Edward O.Wilson

ELSEVIER
NEW YORK · OXFORD · AMSTERDAM

ELSEVIER NORTH-HOLLAND, INC.
52 Vanderbilt Avenue, New York, NY 10017

ELSEVIER SCIENTIFIC PUBLISHING COMPANY
335 Jan Van Galenstraat, P.O. Box 211
Amsterdam, The Netherlands

Library of Congress Cataloging in Publication Data

Barash, David P
  Sociobiology and behavior.

  Bibliography: p.
  Includes index.
  1. Psychology, Comparative. 2. Social psychology.
3. Social behavior in animals. I. Title.
BF671.B18   301.1   76-54359
ISBN 0-444-99029-1
ISBN 0-444-99036-4 pbk.

MANUFACTURED IN THE UNITED STATES OF AMERICA

Designed by Loretta Li

To Jenny and Eva

Contents

Preface

Sociobiology is a whole new way of looking at behavior. It is the application of evolutionary biology to social behavior, an approach that has proven successful in animal studies and that may hold promise for a greater understanding of human behavior as well. This is a new and exciting field; its "classic" formulations and discoveries are literally happening *right now*, nearly all of them the work of biologists. Old concepts are receiving a fresh look and the possibilities for an innovative, multidisciplinary approach to animal and human social behavior appear greater now than ever before. Yet, while social scientists are vaguely aware of these stirrings (and sometimes vaguely uncomfortable because of them!), they have had little opportunity to find out where biology is "at" relative to social behavior. And this is a shame, because biology is in fact knocking right at the front door. I hope this book will help let it in.

I have several goals in writing this book. One is to present the essential themes of sociobiology in a manner that will be both palatable and informative to students in the behavioral sciences who have not had prior exposure to this area—to present the salient facts and theories upon which sociobiology is based. But sociobiology goes further: The basic, organizing principle of sociobiology is evolution by natural selection, a concept that has not received much favorable attention from the behavioral sciences. Accordingly, I have also intentionally assumed something of an advocacy role, seeking to "make a case" for the relevance of evolution to the social behavior of all animals, including *Homo sapiens*. As in a legal brief, facts are presented and arguments developed; this is not purported to be Truth, although out of such proceedings informed judgments will ultimately arise.

In keeping with its intent as a "persuasive primer", this book is directed particularly toward introductory-level courses in psychology, sociology, anthropology, biology, and ethology, where it can help provide a supplemental and perhaps rather novel alternative view to that found in traditional curricula. It assumes no background other than high school biology; in order to keep pages and costs down, much detail is omitted and examples are kept to a minimum. Primers typically do not reference the original literature beyond occasional "suggestions for further reading", perhaps at the end of each chapter. By contrast, I have included fairly extensive bibliographic referencing, largely in support of my hope that this book will serve as an introduction to the rich trove of data and ideas already available in the literature of sociobiology.

Nonetheless, what remains is a bare-bones treatment that hopefully will convey the basic thrust of sociobiology without doing too much violence to its complexity. Because sociobiology is experiencing rapid growth with great possibilities for further advancement, I have indulged rather freely in biologically grounded speculation, in hopes of generating further thought and new directions. Wherever it occurs, I have been at pains to indicate the bases for such speculation and to make it clear whenever I stray from established fact. I believe that such intellectual journeys can be of great value so long as they are recognized for what they are, and I trust that the reader will agree.

This book does *not* provide a comprehensive, balanced view of all social behavior. I have not endeavored to combine present knowledge of developmental, learned, cultural, hormonal, and genetic components of social behavior, providing appropriate weightings for each and coming up with a true synthesis of social behavior in all living things. Possibly, such a thing can be done sometime in the future, but not yet. For now, I hope only to make a case for the sociobiological approach; it is therefore one-sided and "pushes" an evolutionary perspective. This is not to deny the traditional concerns of social science, i.e., the role of environmental contingencies in generating behavior, but, rather, to suggest some alternative or at least additional concerns. It may or may not provoke agreement and possibly even anger; it will be successful it it provokes thought.

Because the present volume is directed mainly toward students in the social sciences, I have emphasized vertebrates, especially birds and mammals, and have restricted mention of invertebrates to consideration of the social insects, whose role in the development of certain theoretical issues, especially kinship theory, remains of central importance. Most of the chapters have been written specifically with nonhuman animals in mind, since the foundations of sociobiology have been established most firmly upon such systems. On the other hand, I have not been reluctant to suggest implications for human behavior, wherever they seemed reasonable and not too "far out". The concluding chapter on human sociobiology is avowedly speculative and perhaps a bit outrageous; I hope it will be accepted in the spirit in which it is offered: a sometimes playful but nonetheless sincere effort to expand our perception of ourselves.

Finally, readers desiring further technical information are urged to consult E. O. Wilson's magnificent, if rather imposing, volume *Sociobiology: The New Synthesis* (1975). Virtually all issues raised in the present book are covered in greater detail in this carefully documented work. Other valuable material can be found in Brown's *The Evolution of Behavior* (1975) and Alcock's *Animal Behavior: An Evolutionary Approach* (1975), as well as in textbooks of evolutionary ecology such as those by Emlen (1972), Pianka (1974) and Ricklefs (1973). In addition, several disciplines relevant to sociobiology are not considered directly in the present book; this omission is not a judgement upon the merit of these disciplines, but a reflection of brevity's requirements. These additional areas and recommended sources for further information include the following: communication (Otte, 1974), foraging strategies (Schoener, 1971), behavioral ontogeny (Moltz, 1971), and population biology (Wilson and

Bossert, 1971). Enjoy yourselves, but be quick about it: So much is happening right now that all the above references, and this volume as well, will probably be obsolete very soon. I certainly hope so.

Seattle, 1976

I would especially like to thank Bob Lockard and Pierre van den Berghe, both of whom read through the manuscript and made helpful suggestions. I owe an enormous debt to all those cited in the References, and doubtless to many others who were not. Finally, my appreciation to the staff at Elsevier, notably Bill Gum who recognized merit in an early draft, Ethel Langlois who appreciated the value of a well-illustrated book and Margaret Quinlin who assisted in various aspects of the labor and delivery.

Foreword

Sociobiology and Behavior is swift and light in style, but it is entirely serious in content. Professor Barash has provided an excellent primer that can serve well in biology courses but is also particularly suitable as an introduction of the subject to social scientists. Sociobiology, he makes clear, is the child of ethology and population biology. The union of these two disciplines, which themselves are experiencing rapid growth, has produced some surprising results. Not only has new light been shed on the mechanisms of social behavior, but the *structure* of societies has been laid open to biological analysis for the first time.

Animal societies are now being compared in a way that yields general principles. Sociobiology, by its definition as the systematic study of the biological basis of all forms of social behavior in all kinds of organisms, is above all a comparative science. From the biologist's point of view, to put the matter as starkly as possible, the social sciences are concerned with only a single mammalian species. The biologist makes use of the fact that we live on a planet of staggering organic diversity. About one million species of animals have been described; various estimates place the actual number alive between three and ten million. Sociobiology derives its information from the comparison of the tens of thousands of these forms that have evolved some grade of social life. Each species represents an evolutionary experiment that can be consulted in great depth, with ramifications that reach into ecology, genetics, and physiology.

The purpose of sociobiology is not to make crude comparisons between animal species or between animals and men, for example simply to compare wolf and human aggression or slavery in ants as opposed to that in men. Its purpose is to develop general laws of the evolution and biology of social behavior, which might then be extended in a disinterested manner to the study of human beings. In the same way that biologists have learned about heredity from the study of fruit flies and the little bacterium *E. coli* and applied the principles of human heredity, we expect to extend such general principles of sociobiology as can be devised to assist in the explanation of human behavior.

The advent of human sociobiology is not central to the new discipline but it is clearly the most problematic and generally interesting development to emerge from the new discipline, and it deserves a special comment. The evidence is very strong that there does exist a human biogram, a pattern of potentials built into the heredity of the species as a whole. In some cases we vary very little. The facial expressions used to express the basic emotional expressions appear to be the same across all cultures. Other transcultural traits include the incest taboo, the use of bodily ornaments in multiple forms of social display, and the construction of elaborate kinship rules. *Homo sapiens* shares with other social mammals a tendency toward male dominance systems, a sexual division of labor, prolonged maternal care, and an extended socialization of the young based in good part on social play. It is vital not to misconstrue the political implications of such generalizations. To devise a naturalistic description of human social behavior is to note a set of facts for further investigation, not to pass a value judgment or to deny that a great deal of the behavior can be deliberately changed if individual societies so wish.

In other traits human beings are of course much more variable, most notably in the details of language, technology, and fashion. Human behavior is dominated by culture in the sense that the greater part, perhaps all, of the variation between societies is based on differences in cultural experience. But this is not to say that human beings are infinitely plastic. Even during periods of relative isolation human societies did not drift apart in the manner of stars in an expanding universe. When compared with the thousands of other social species, human beings can be seen to operate within limits. They are social in a way that is basically mammalian yet distinctive among the mammalian species. The human constraints appear to have evolved from a mammalian plan over the past several millions of years, during which our ancestors lived in a hunter-gatherer economy.

To understand that evolutionary history and the contemporary biogram that it produced is to understand in a deeper manner the construction of human nature, to learn what we really are and not just what we hope we are, as viewed through the various prisms of our mythologies. Assisted by sociobiological analyses, a stronger social science might develop. An exciting collaboration between biologists and social scientists appears to have begun.

EDWARD O. WILSON

Cambridge, Massachusetts
September 18, 1976

Sociobiology and Behavior

O N E

The Domain of Sociobiology

If we take an objective view of the study of behavior, particularly as it has been conducted by the social sciences, psychology, sociology, and anthropology, we would have to agree that it has lacked a coherent, underlying framework—what Kuhn (1970) has described as a "paradigm". A paradigm constitutes a basic interpretive structure that serves to organize our inputs from the natural world. It provides unity to our experiences, and it interprets the data of our science. A paradigm may also go beyond the structuring of information already received, that is, it may selectively permit the reception of certain information and cause us to screen out data that do not "fit". Like the ever-present gate-keepers of mythology, it may also restrict our world view and, without our consciously realizing it, it may even set limits on the phenomena of which we are aware. But even this process, while overtly restrictive, is creative. Thus, a good paradigm encompasses what is known, suggesting directions for future research. But, at the same time, it should not be so complete as to explain everything, therefore illuminating nothing. It should be useful, adjustable to a point, and, like any hypothesis, susceptible to refutation.

Thought processes in the Western world operate under certain grand paradigms; for example, cause and effect; the assumption that the information we receive from our senses is an accurate representation of the "real" world; and the assumption

1

of continuity of the world in time and space. Many of our paradigms are so pervasive that it is an effort to step back and recognize them. On a more empirical level, atomic theory has proved to be a valuable paradigm: it has successfully replaced alchemy and "earth, air, fire, and water". Newtonian mechanics had provided a foundation for modern physics that has been adjusted somewhat during this century by the new and somewhat parallel view of Einsteinian relativity. Similarly, the earth-centered Ptolomeic universe was unable to account for a growing body of data; this insufficiency spawned a paradigmatic revolution culminating in the Copernican view. Textbooks, such as Pauling's *General Chemistry* and Feynman's *Lectures on Physics*, reveal the power of their paradigms in the intellectual momentum with which the premises of each field are unfolded, checked against the data, and then used to generate further premises, etc.

Biology has an equally cogent paradigm: evolution by natural selection. Thus, evolutionary theory provides a unifying interpretive framework, as revealed in such textbooks as Keeton's *Biological Science*, Simpson and Beck's *Life* and Orians' *The Study of Life*. I propose that evolutionary theory may also constitute a valuable paradigm for all of the life sciences and especially for the study of animal behavior, both human and nonhuman. This is the heart of *sociobiology*: the application of evolutionary biology to the social behavior of animals, including *Homo sapiens*.

Lacking an adequate paradigm, students of animal behavior in the twentieth century have resembled nothing so much as a gaggle of novitiate monks, rushing to sit at the feet of one *guru* after another, eagerly seeking some form of generalized Enlightenment. But they haven't quite "made it". Jacques Loeb's (1906) theory of forced movements or tropisms (later, taxes; Fraenkel and Gunn, 1940) attracted followers for a time. There was excitement over Sherrington's identification of the reflex arc (1906) and further enthusiasm following the discovery that, under certain conditions, some reflexes could be conditioned to occur in association with external stimuli: the name Pavlov should ring a bell.

2

Then came the great age of learning theorists in psychology: Guthrie (1935), Hull (1943), and Spence (1960). Perhaps the most influential theorist has been B. F. Skinner (1938), although he would probably disavow this description, preferring instead to rely on the "atheoretic" analysis of behavioral data. To Skinnerians, *behavior* is an appropriate study whereas *mind* is not. But there is also an attempt at a paradigm here: Reinforcement theory (behavioral modification) is based on the underlying premise that behavior is influenced by its consequences. This approach characterizes much research today, particularly in psychology, and it has proven to be useful in manipulating behavior. Thus, the basic principles of operant conditioning are effective in modifying the behavior of animals, including humans. The spectacular animal feats of the Moscow circus are testimony to the effectiveness of this approach. But even in its manipulative function, problems have arisen as animals persist in behaving in ways that are outside the traditional paradigms of learning theory (Breland and Breland, 1961; Bolles, 1970; Garcia et al., 1974).

Significantly, most limitations to learning theory are all rendered explicable by the addition of an evolutionary perspective. Beyond this, theories of behavior modification and social learning often provide little insight into the natural behavior of free-living animals. They do not help in understanding why scout bees dance about in funny ways after discovering a food source; why wolves live in packs; and why robins sing every spring. Similarly, learning theory suggests how to design teaching machines and slot machines, and how to toilet train a child, but it may be of limited value in elucidating the complex patterns of human behavior.

Consider a simple detour experiment where an animal must go away from a goal in order to reach it eventually (FIG. 1.1). Dogs are no good at such tasks: a dog strains at its leash; whines pitifully; looks at you with its sad brown eyes; runs about wildly; and finally may fall asleep. Then it starts again. Eventually, by luck alone, it might find itself on the far side of the intervening post, whereupon it rushes triumphantly to the food. By contrast, common tree squirrels are uncommonly good at solving

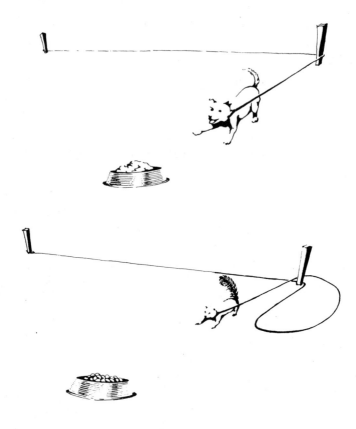

FIGURE 1.1. DOGS VERSUS SQUIRRELS ATTEMPTING A DE-TOUR PROBLEM. (A) A dog does poorly, straining at its leash. (B) A squirrel solves this problem readily, it goes around the post and gets the food. (ILLUSTRATION BY D. COOK)

this problem. After surveying the situation, the animal proceeds surely to circumnavigate the post and get the food.

Why the difference? This type of result is puzzling to learning theorists, especially since squirrels are universally recognized as less intelligent than dogs. Even their brains are obviously simpler, with fewer convolutions in the cerebral cortex. The answer is actually straightforward if we consider the natural history of squirrels and dogs. Dogs live in a two-dimensional world: If they want something, they go and get it. Squirrels, on the other hand, live in trees. In their three-dimensional world, a squirrel wanting to go from tree to tree has a choice: it can descend the trunk; go along the ground and then climb the tree; or it can remain in the treetops, seeking a place where branches of two adjacent trees are in contact. The former strategy would expose the squirrel to its ground-dwelling predators; the latter would therefore be safer, but it would often require the ability to go initially away from the goal so as to achieve it eventually. In other words, the ancestors of present-day squirrels were good at solving detour problems. Rephrasing this more accurately in the language of evolutionary biology: ancestral squirrels that were relatively more adept at solving such problems left more offspring than did those who were less well endowed. Thus, the ability to conduct successful detours was favored by natural selection, such that each population of squirrels came to be composed of individuals who were good at going away from goals in order to reach them.

Numerous examples of this sort can be adduced, and yet, for the most part, behavioral science has remained wedded to the same inadequate paradigm—the modification of behavior by experience. Even when animals are studied, the behaviors are almost unrelated to natural situations, and very few species are examined. In fact, comparative psychology has become the study of maze running and bar pressing by white rats. Psychologists have been urged to consider other species (Beach, 1950) and to incorporate evolutionary thinking into their science (Lockard, 1971). But in general they have not done so. Comparative psychology today is characterized by a wealth of sophisticated gadgetry, elaborate data-gathering devices, and, often, elegant statistical analyses. But, it may well be missing the forest for the trees.

Sufis are an ancient sect of Moslem mystics who have been particularly adept at highlighting the human penchant for self-delusion. When Nasrudin, the idiot-savant of Sufi literature, was grovelling in the dusty street, a friend asked what the problem was. Nasrudin explained that he had lost his house key and was searching for it. The friend offered to help and after a long, unsuccessful search he asked whether Nasrudin was certain that the key was dropped in that spot. "Oh, no", replied Nasrudin, "I dropped it near my house", a full block away. Seeing his friends perplexity, Nasrudin explained, "But the light is so much better here"!

There may well be a lesson here: if psychology and the other behavioral sciences truly wish to find a key, they may have to look where it is a bit darker. Actually, considerable movement has already occurred, especially with recognition of "biological boundaries of learning" and similar phenomena (Shettleworth, 1972; Seligman and Hager, 1972; Hinde and Stevenson-Hinde, 1973). It may be that social psychology, sociology, and anthropology will also move increasingly toward sociobiology. If so, these disciplines will bring much light with them, and it will be appreciated!

Ethology is the biological study of animal behavior. As such, its orientation has been evolutionary from the start, with primary attention directed toward evaluating and understanding the diversity of behaviors in free-living animals. Partly in reaction to the extreme environmental determinism of early American psychology, ethology emphasized species-specific and genetically mediated behaviors. But although the discipline was evolutionary, it tended to emphasize a strictly historical approach, particularly the identification of behavioral homologies and phylogenies based on the proposition that behavior has evolved just like structure (Heinroth, 1910; Lorenz, 1950). This, in itself, was a major contribution, along with recognition of the value of observing each behavior in its entirety, in the environment within which it evolved. For a modern treatment of this view of behavior, see Eibl-Eibesfeldt (1975).

But classic European ethology fell prey to the *Rumplestilskin effect* (Smith, 1975), the beguiling notion that if we name something, it will go away. Thus, ethology generated such phrases as "fixed action pattern", "innate releasing mechanism", and "action specific energy". These notions may be of considerable heuristic value—i.e., they help identify and order phenomena—but they don't really *explain* very much. Worse, ethology was for the most part only quasi-evolutionary, employing a historic or, at best, a static view of the evolutionary process.

Whereas psychology has operated in large part independently of the ideas of evolution, sociology and anthropology

6

have had a curious love-hate relationship with Darwinism. Thus, by the latter half of the nineteenth century, evolution by natural selection was becoming increasingly acknowledged as fact. Unfortunately, the mechanism was commonly misinterpreted. Evolution was thought (incorrectly) to operate by aggressive, life-and-death competition among individuals in which the "more fit" subjugated the "less fit". This was the era of laissez faire capitalism in the Western world, and the captains of industry eagerly embraced this view of Darwinism as a justification for the social inequities prevalent at the time. "Social Darwinism" therefore flourished, with its reassuring message that social and economic exploitation simply represented the inevitable unfolding of nature's way. Sociologists and anthropologists contributed to this travesty, much to their later chagrin, and in fact they subsequently did penance for their sins by proclaiming that human beings enjoy absolutely unlimited behavioral potential, as *tabula rasa* (blank slates) upon which experience can write as it will. So, after a brief fling with evolution, sociology and anthropology embraced a conception of human behavior that attributed everything to early experience, socialization, cultural norms, etc. In a sense, this rampant environmentalism is a paradigm in itself, but despite a plethora of social theories both fields have a disturbing lack of intellectual cohesion. Recently, efforts have been made to incorporate evolutionary thinking into anthropology (e.g. Tiger and Fox, 1966), sociology (van den Berghe, 1974), and social psychology (Campbell, 1975), but such attempts are very much a minority.

Thus, evolution has received little attention from both social scientists and most biologists concerned with behavior. To the lay person, evolution has something to do with dinosaurs, and human beings evolving from the apes. Even to most biologists, it has been merely a historical document, an explanation of how things got to be the way they are. Evolution is generally recognized as having produced the arena within which behavior occurs. Indeed, it is responsible for the participants as well. But, like religious hypocrites, most behavioral scientists have been content to profess the faith, while functioning on a daily basis as though the organizing constructs are somehow irrelevant. Or,

7

like the "God is dead" movement: evolution may have originated things, but it isn't relevant to the actual, present-day workings of the world. Indeed, such religious imagery is appropriate here, since evolutionary theory can easily take on the trappings of dogma, whereupon it deserves the criticisms already leveled against European ethology, substituting words for understanding and stifling research on legitimate, tractable questions (Lehrman, 1953). The military strategist, von Clausewitz, said that "he who would defend everything, defends nothing". Similarly, to explain everything is to explain nothing, just as the attribution of natural phenomena to divine intervention has not proven to be a productive scientific approach. But, evolution is relevant to living things: it not only is the way life came to be and the mechanism that will be responsible for changes in its future but it also is the guiding principle that makes sense of why living things are as they are.

Sociobiology has already provided enormous insights into the behavior of nonhuman animals, and accordingly there is every reason to extol its virtues to students of animal behavior. There is also reason for optimism concerning its value when applied to human behavior, although in this regard proselytizing for open-mindedness seems more appropriate than does single-minded adherence to evolution as the only paradigm worth pursuing. In any case, sociobiology has emerged from the recognition that behavior, even complex social behavior, has evolved and is adaptive. Its excitement derives from the further recognition that evolution has a great deal to say about behavior; it is the underlying thread that unifies all living things, not only in terms of genealogical relatedness and, therefore, ultimate unity but also as the primary mechanism to which all life is subject. Correctly used, evolutionary theory is a predictive and analytic tool of enormous power. The strength of sociobiology derives from its grounding in the universalities of evolutionary biology. Its promise for the study of behavior lies in the hope of a good paradigm.

T W O

How It Works:
Evolution as a Process

Most people think they understand evolution; few do. Survival of the fittest comes readily to mind, but the phrase itself does not provide much insight and may be misleading. An understanding of evolution by natural selection is essential here, because evolutionary theory provides the basic paradigm for sociobiology. Accordingly, in this chapter I will outline the theory, with emphasis on possible areas of misinterpretation. I will not be concerned here with providing evidence for the legitimacy of evolution itself; such material is available from many sources (e.g., Simpson, 1949; Weller, 1969; Moody, 1970).

An important distinction ought to be made at the outset: there is *evolutionary theory* and there are *theories of evolution*. The former is not *just* a theory; it probably is as close to truth as we can get in the natural sciences, analogous to the atomic theory or the theory of relativity. On the other hand, although there is near-universal agreement as to the general process, there are numerous *theories* of evolution, squabbles over some of the details. The view presented here is the one most generally acknowledged: the modern synthetic theory of evolution (Huxley, 1942) comprising a synthesis of natural history, logic, and genetics.

Darwin's great contribution was not the *discovery* of evolution. The notion that all living things had evolved from a common origin was around for a long time (see Eiseley, 1958 for a won-

9

derfully literate history of evolutionary thought). Rather, Darwin described the *mechanism* by which evolution operates and then buttressed this description with an overwhelming array of supporting facts. The mechanism is natural selection and it is the key to the evolutionary process.

Basically, Darwin's reasoning was:

1. All living things have a tendency to overproduce. If a population is to remain the same size, two parents must produce two successful offspring during their lifetime. If they produce more than two offspring, the population will increase and, because each offspring can itself become a parent, the increase can become explosive (exponential). If the parents produce fewer than two offspring, the population will decline and eventually will become extinct. Obviously, animals tend to produce more than two offspring: a codfish produces a million or more eggs at a single spawning; a robin lays four eggs at a time. Furthermore, a robin may produce several clutches before it dies. Unchecked, such profligate reproduction would eventually blanket the earth with a writhing and ever-growing mass of flies, oysters, codfish, robins, elephants, blue whales, or whatever species we choose to consider.

2. Despite each species' persistent capacity to overproduce, populations tend to remain remarkably stable from one generation to the next. To be sure, the numbers of some animals, such as lemmings and snowshoe hares, tend to fluctuate somewhat, but generally within limits. Over long periods of time, relative stability is the rule.

3. Individuals differ, and, to some extent, those differences are passed on to their offspring. Thus, among sexually reproducing species, only identical twins are identical; otherwise, differences exist among individuals. They may be slight, but they are real.

4. Given stability of the numbers in the face of potential profligacy and exponential increase, it is apparent that some individuals are more successful than others in producing offspring and/or some offspring are more success-

ful than others in becoming adults. In other words, competition occurs. Survival of the fittest simply means that those individuals possessing characteristics that render them more capable of surviving and reproducing will do so and will be more successful; that is, they will be better represented in the next generation than will those individuals who are less fit. This is natural selection: the differential reproduction of individuals within a species, from one generation to the next. Natural selection follows irresistably from the facts of (1) each species' potential for overproduction; (2) the absence of such overproduction; (3) the existence of differences among individuals; and (4) the resulting competition among individuals, as a result of which some are more successful than others.

5. Finally, the process of natural selection, operating by the differential representation of each individual's offspring in succeeding generations, produces a gradual change in the make-up of each species. This change is evolution.

Of course, natural selection can lead to evolutionary change only when the differences between individuals reflect some degree of underlying genetic differences. If all individuals are genetically identical, then it does not matter which ones are producing more successful offspring: the characteristics of the next (offspring) generation will not differ from those of the earlier (parental) generation. This can be demonstrated by graphing the distribution of traits in any natural population. Although a variety of statistical distributions are possible, the most common pattern approximates a normal or bell-shaped curve (FIG. 2.1). Other curves may be obtained, depending upon the trait in question; the exact shape of the distribution does not affect the discussion here, so a normal curve will be assumed for the sake of simplicity. An intermediate condition is generally most common within any population, both extremes being present with increasing rarity. For a human example, the distributions of heights, weights, physical strength, and IQ fit this pattern.

Now assume that the variation in any trait is due entirely to

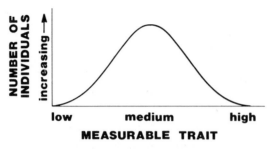

FIGURE 2.1. A COMMON DISTRIBUTION FOR
ANY MEASURABLE TRAIT IN A NORMAL POPULA-
TION. In this normal or bell-shaped distribution,
there are few individuals at the low level of the
trait and equally few at the high level. The largest
number of individuals shows the trait at the
medium level.

environmental differences experienced by different individuals; i.e., there is *no* genetic difference with regard to the trait
being considered. In this case, it would make no difference
whether individuals at either end of the distribution, in the
middle, or anywhere else were selected as parents to produce
the next generation; the distribution of traits among their offspring would not differ from the initial population, regardless
of who the parents were. For example, consider human height
and make the (untrue) assumption that it has no genetic component, but, rather, it is entirely a function of nutrition. In this
case, we could increase human height by improving nutrition
but not by selective breeding: if nutrition or other environmental factors alone determine height, tall parents would be no
more likely to produce tall children than would short parents.

Natural selection is differential reproduction. According to
theory, evolutionary change is brought about by differential
reproduction. It should therefore be possible to produce an evolutionary change by causing individuals of one type to reproduce (differentially) more than others. Actually, plant and animal breeders have been doing this for a long time. In a now-
classic experiment conducted at the turn of the century, the
botanist Johanssen attempted to select artificially for large and
small beans in standard garden plants. He observed a normal

distribution of bean sizes in the parent population, then chose individuals from the large end of the distribution and cross-fertilized them, doing the same with individuals from the small end. He continued this procedure for several generations, expecting that eventually an evolutionary change would occur, i.e., the distribution of the traits would become noticeably different among the offspring of the selected lines. This did not occur (FIG. 2.2). Understandably, these findings were taken as evidence that natural selection could not produce evolutionary change. However, it happened that he had chosen plants (beans) that were highly inbred and homozygous.* In other words, there was essentially no genetic variability in the original population. The difference between the large and small beans was due entirely to environmental factors such as the amount of light and water they received or whether they grew in crowded conditions.

Rather than demonstrating a failure of evolution, Johanssen's experiment highlighted an important distinction in evolutionary biology, that between *phenotype* and *genotype*. An organism's phenotype is any actual, directly observable characteristics of an organism—such things as size, color, or shape and anything we can measure or count, from structure to behavior. Figures 2.1 and 2.2 show distributions of phenotypes. By contrast, the genotype is the genetic makeup of an organism. It is discernible only by its influence on the phenotype but is distinct from it. In Johanssen's example, there was phenotypic diversity without genotypic diversity. Natural selection will produce evolutionary change only if the difference between the phenotypes of those individuals that are reproductively favored (selected) and those that are not, is due to differences in the genotypes between individuals of these two groups.

Since Johanssen's early work, numerous studies have confirmed the competence of selection to produce evolutionary change, given only some degree of genetic variation underlying the observed phenotypic variation (FIG. 2.3). The basic procedure is to choose as parents individuals who differ in some way

---

* A term indicating that the genes responsible for the trait in question, seed size, were identical among all individuals in the population.

13

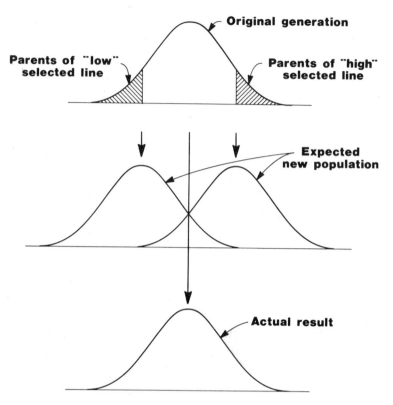

FIGURE 2.2. EXPECTED VERSUS ACTUAL RESULTS WHEN GENETIC VARIATION IS ABSENT. When the distribution of a trait in a population simply reflects environmental differences experienced by the different individuals, their selection cannot produce evolutionary change.

from the mean (average) of the original population. Their offspring will comprise a population with a somewhat different distribution of the phenotype in question, reflecting in part a different genotypic distribution. This shift is an evolutionary change. When it occurs as a consequence of human decisions as to which individuals shall be reproductively favored, it is *artificial selection*. This is the essence of scientific plant and animal breeding. When nature makes the reproductive decisions, it is *natural selection*.

When natural selection consistently favors individuals removed in the same direction from the mean, there will be a progressive change in the distribution of the trait in succeeding

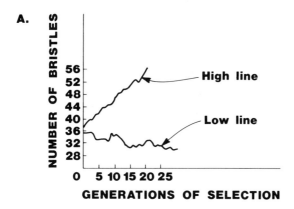

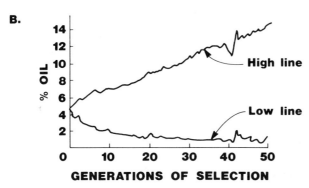

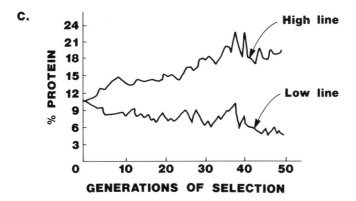

FIGURE 2.3. EXAMPLES OF EVOLUTIONARY CHANGE PRO-
DUCED BY ARTIFICIAL SELECTION. (A) Selection of high versus
low number of abdominal bristles in fruit flies. (*Modified slightly
from Mather and Harrison, 1949.*) (B) Selection for high versus
low oil content in corn. (C) Selection for high versus low pro-
tein content in corn. (*Parts* (B) *and* (C) *modified slightly from
Woodworth et al., 1952.*)

15

generations, moving in the direction of those individuals that have been selected. This is *directional selection*, the most commonly identified type of evolutionary process (FIG. 2.4). When most people think of evolution, they assume directional selection. For example, the ancestors of *Homo sapiens* experienced a progressive increase in brain size because individuals with larger brains, and some genetic basis for this trait, left more offspring than those with smaller brains. Brain size presumably contributed to reproductive success, perhaps by improving the ability to manufacture and use tools, or by improving communication within hunting bands and/or between parents and children, etc. Similarly, during the Industrial Revolution in England, an increase in the frequency of dark-colored moths and a decrease in the frequency of light forms occurred. This change came about because the deposition of soot on trees rendered the light forms more conspicuous to predators (birds)

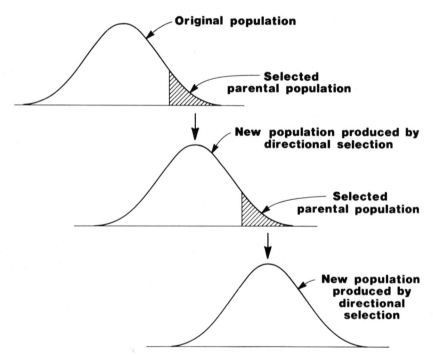

FIGURE 2.4. DIRECTIONAL SELECTION. The distribution of a trait in a population changes from one generation to another, with the mean shifting in the direction indicated by the selected parental population in each case.

than the dark forms, thus selecting for the dark form relative to the light (Kettlewell, 1956; 1957; 1958).

By contrast, natural selection may also operate upon a population without changing the mean genotype or phenotype. Thus, whereas directional selection involves differential reproduction by individuals at one extreme of a distribution, it is also possible that both extremes are at a disadvantage and are selected against relative to those at the mean. This selection against the deviants is known as *normalizing* or *stabilizing selection*; it is probably the most common form of selection in nature but is rarely noticed because of its unspectacular effect, the *prevention* of change. In fact, stabilizing selection is a major reason for the mean of a population being where it is. Under a regime of stabilizing selection (FIG. 2.5) the mean remains constant across generations while the variance of the population, a statistical measure of the spread of the distribution, decreases.

In an important demonstration of stabilizing selection, H. C. Bumpus collected 136 sparrows that had been stunned by a

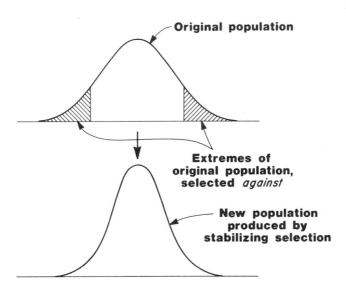

FIGURE 2.5. STABILIZING SELECTION. The average distribution of a trait in a population does not change from one generation to another. Selection *against* the extremes produces a narrower distribution (less scatter on either side of the mean) but no change in the mean itself.

passing hurricane. He obtained various body measurements on each sparrow, and he found that 64 members of the original population died from injuries and 72 survived. The surviving population was composed of individuals whose body measurements were significantly less extreme than were the analogous measurements of the unsuccessful population. Apparently, wings that are too long can provide too much sail in a strong wind, while the opposite extreme may have insufficient propulsion. Natural selection, by operating against such deviants, was stabilizing the physical characteristics of the sparrow population at an ecological optimum.

Finally, it is possible for selection to operate against the mean, actually favoring both extremes. The resulting *disruptive selection* produces a multimodal distribution (FIG. 2.6). However, this pattern is probably rare and not particularly important in nature. Closely related to disruptive selection is *frequency-dependent selection*, in which rare types enjoy an advantage only while they are rare. As they increase in frequency because of their advantage, this advantage wanes and they decrease in abundance. The resulting equilibrium seems to be characteristic of certain prey species in which predators home in on abundant types and ignore rare ones (Moment, 1962) and also in mate-selection behavior in which rare types are often favored, so long as they are not pathological (Petit and Ehrman, 1969).

It should be emphasized that the three primary types of selection, directional, stabilizing, and disruptive, are identical insofar as the underlying evolutionary process is concerned. All of them come from differential reproduction of individuals. The only differences lie in which individuals are selected and, therefore, what consequences arise for the distribution of the trait in question. The environment is the ultimate natural selector: given a particular environmental regime that is stable over many generations, the distribution of traits within a population will tend to equilibrate through stabilizing selection. If the environment changes, conferring reproductive advantage upon individuals that are somehow removed from the mean, then by definition directional selection will be occurring. When the en-

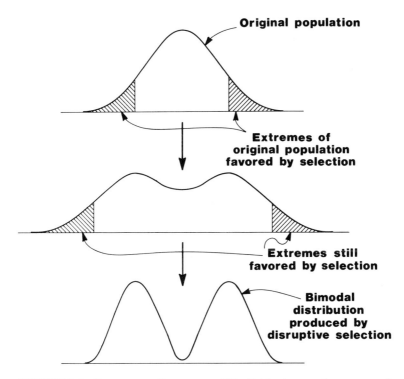

FIGURE 2.6. **DISRUPTIVE SELECTION.** If individuals at each extreme of a distribution are selected relative to those closer to the original mean, then directional selection occurs in two directions at the same time. The result is a multimodal distribution.

vironment stabilizes once again, the selected population will stabilize, again under the influence of stabilizing selection.

Knowing that selection often acts against the extremes, producing a narrower distribution with reduced variability in a population, it seems paradoxical that considerable variability is apparent in each generation, only to be winnowed anew by selection. However, we must remember that in the perspective of evolutionary time (millions of years) all environments change. Such changes alter the adaptive value of traits, conferring reproductive disadvantage on those previously favored and benefitting others who are distributed away from the mean. Accordingly, there is an advantage to "hedging one's bets," i.e., producing offspring of sufficient diversity to remain in the evolutionary game, even if the rules are somewhat changed. Many species show a variety of characteristics whose effect is to store genetic diversity. Perhaps the best known of these is *heterosis* or heterozygote superiority: if the heterozygote (say, genotype *Aa*) is reproductively superior to either homozygote (*AA* and *aa*), then neither gene *A* nor gene *a* can be eliminated by selection alone, because cross-matings of *Aa* individuals will maintain both genes in the population. In fact, even if both *AA* and *aa* genotypes are lethal (and therefore strongly selected against), they will continue to pop up in each generation simply because of the way the genetic system works.

One of the most important examples of heterosis in human populations is sickle-cell anemia. Among certain West African populations, the gene for sickle-cell (Hb_s) is present in surprisingly large numbers. Given that individuals who are homozygous for the sickle-cell gene ($Hb_s Hb_s$) do not survive to reproduce, we would expect that the gene would be selected against quite strongly, and should therefore eventually disappear. And yet, the sickle-cell gene occasionally represents as much as 40% of the total, with normal hemoglobin (Hb_N) comprising only 60%! The reason is that heterozygous individuals—those carrying one normal gene and one sickle-cell gene ($Hb_N Hb_s$)—are less susceptible to malaria than are normal individuals ($Hb_N Hb_N$). As a result, these "carriers" are reproductively more successful than are normal individuals, and each time they reproduce, the sickle-cell gene is retained in the population, despite its disadvantages in the homozygous state. In this manner, genetic variability is retained in natural populations, not through some mystical, prescient force acting for the good of the species, but rather, through selection favoring genotypes that are reproductively successful. Of course, with environmental change (the habitation of areas such as North America where malaria is much less of a selective agent), the adaptive advantage of this particular trait is essentially eliminated.

Such discontinuous, genetically mediated forms, *polymorphisms*, can also be maintained by discrete environments, each with different selective regimes favoring different forms, or by seasonal variations in selective regimes within the same environment (see Sheppard, 1960, for an excellent introduction to polymorphisms, their maintenance, and their significance). Heterosis and environmental variation therefore increase genetic variation. However, it seems unlikely that they occur to serve this function; rather, the maintenance and preservation of genetic variation is best viewed as an incidental *result* of natural selection than as a *cause* of any particular pattern.

Thus, once genetic variation occurs in a population, it may be retained in a variety of ways. But, how is it produced in the first place? For the answer, we must turn briefly to mutation and recombination, two biological processes of the greatest importance. Evolutionary change not only requires some correlation between phenotype and genotype; it also requires a range of variation for both so that selection has the raw materials upon which it can operate. *Mutation* and *recombination* provide the ultimate sources of genetic diversity.

Of the two, mutation is more essential. Broadly speaking, a mutation is any heritable change in a gene or a chromosome. These changes are errors, mistakes that occur during the normal process of reduction division in the cell (meiosis) when eggs or sperm are being produced. Thus, a part of the genetic material (DNA) in the cell may fail to replicate exactly, causing a gene mutation. Or, whole chromosomes may break or be represented abnormally in the gametes produced. Both gene and chromosomal mutations are passed on to the offspring that are produced, who will differ somewhat by virtue of possessing them. The vast majority of mutations are harmful; this makes sense if we consider that living things represent complex, highly nonrandom combinations of molecules and cells, produced by the differential reproductive success of those combinations that are best suited to their particular environments. A mutation introduces some randomness into this highly ordered system. Accordingly, it will almost always make the system function less well. Rarely, however, a mutation may actually cause an improvement and when it does it will be selected, naturally. In other words, individuals possessing this adaptive mutation will leave more offspring than will individuals whose phenotypes are different because their genotypes lack the mutation in question.

Many people have a "horror movie" conception of mutations, in which mutant human beings, for example, may appear as hairy, one-eyed monsters with three arms. In fact, most mutations produce small, barely discernable effects, and almost certainly the likelihood that any mutation will be beneficial and therefore selected varies inversely with the magnitude of its phenotypic effect. This is one reason why evolutionary change is invariably slow. It is ironic perhaps, but all biological progress is based upon mistakes, albeit small ones.

Ultimately, mutations are responsible for all genetic diversity. But from one generation to the next, most diversity is generated by sexual recombination, the reshuffling of chromosomes that occurs every time eggs and sperms are produced, along with the novel combinations that occur when egg and sperm unite to make a new individual. In fact, this is the "reason" for sexual reproduction (see Chapter 6).

In a sense, natural selection is a simple, straightforward process. However, it has led to considerable misunderstanding and the next few pages will be concerned with clarifying points that are often confused. For example, Darwin's expression, "survival of the fittest", has caused a great deal of unnecessary difficulty. It was erroneously equated with violence, aggression, an unbridled world of bloody competition, hence, Tennyson's unfelicitous phrase, "nature red in tooth and claw". Actually, natural selection is considerably tamer than all that. Success in fighting may contribute to evolutionary success but only insofar as it leads to a reproductive advantage. For example, the highly aggressive bull elk who defeats all comers in head-on conflict will actually be selected *against* relative to the individual who fights less and reproduces more. Of course, fighting prowess *may* lead to more reproducing, and reproduction is what really counts. In the world of evolutionary biology, successful reproduction is the ultimate end; fighting is but one of many means to that end.

Natural selection accordingly operates on many characteristics of living things. Organisms may actively compete for status in a dominance hierarchy, ownership of territories, the approval and/or possession of females, access to food or resting sites, etc. But ultimately these factors are evaluated only in terms of their contribution to reproductive success. To a large extent, selection also operates via indirect competition among individuals, for example, for the efficient consumption and utilization of food, whether animal or vegetable; resistance to environmental stress, drought, famine, wind, rain, or predators; and speed of effective growth and development. Some competition may occur with regard to the total number of eggs laid or young born, but the ultimate criterion is the number of *successful* descendants eventually accounted for. In this sense, a smaller number of well-endowed offspring may lead to greater evolutionary success than a large number of relative incompetents. Again, many factors will contribute to scoring points in the evolutionary ballgame, but it is only the final score that counts, not how the game is played, and that is measured in units of differential reproduction.

It is sometimes assumed (incorrectly) that evolution proceeds by dramatic, cataclysmic environmental changes. Indeed, such events may contribute to natural selection, but, they are not essential. Similarly, gross genetically mediated deformities are often thought to be fundamental to evolution, whereas in fact mutations are more likely to be relatively small in their effect, and certainly the likelihood that a mutation will increase the reproductive success of its carrier varies inversely with the magnitude of the phenotypic effect it produces. Thus, evolution by natural selection proceeds by the gradual accretion of genetic characteristics that confer relatively more reproductive success upon each individual in each generation, given the environments that exist in each case. Incidentally, environments in this sense should be defined as broadly as possible, including not only climate but also all possible interactions with predators, prey, competitors, mates, offspring, etc. We should even include the genetic environment of the species itself.

As a consequence of some consistent environmental trend, there may well be a consistent evolutionary response that is the result of directional selection. For example, increased dryness during the Pliocene era may have caused tropical forests in East Africa to recede and be replaced by the more open savannahs of today. This new environmental regime may in turn have selected for bipedalism (upright stance) among the primitive ancestors of *Homo sapiens*, because it permitted greater visibility over the tall grasses, freed the arms for using weapons and tools with which to compete with the large savannah-dwelling predators, etc. At the same time, the need for precise coordination among individuals may have led to the evolution of speech and larger brains. The human fossil record certainly reveals a progression toward upright posture and increased brain size, but the fact of this progression should not be interpreted as a *cause*. It is not scientifically justified to attribute any evolutionary change to some form of historical momentum; this discredited notion is known as orthogenesis in the older literature.

Occasionally, one hears the argument that because natural selection is a random process, it cannot create anything new. The production of variability is indeed random; natural selection is not. In fact, natural selection may profitably be viewed as a device for the generation of highly nonrandom systems. Thus, by selectively evaluating the reproductive success of each individual out of an enormous array of forms, selection introduces order into chaos and produces a nonrandom system, much as the legal expertise represented by the United States Supreme Court greatly exceeds that found in any nine adults chosen at random. Supreme Court justices themselves were *not* created by Presidential selection and Senate confirmation, but the highly nonrandom system we call the Supreme Court *was* created by this process. In a similar sense, all living things are significant departures from randomness, i.e., they are more organized, or represent less entropy, than most of the nonliving world.

Beyond this, selection produces further novelty. By manipulating the likelihood that certain individuals will leave successful offspring that will in turn be more likely to breed with other, also successful individuals, selection greatly increases the probability that certain new genetic combinations will be produced. For example, the probability that ten mutations, each

24

occurring with a frequency of one in a million, will all be present in an individual, is one in 10^{60}. This is an incredibly small number, of the same order of magnitude as the probability of choosing, blindfolded, a single previously designated electron out of all the electrons in the universe. Yet, such an individual could easily be produced after only a few dozen generations of selection. This is creativity of the highest sort. In one sense, of course, nothing is really novel: matter can be neither created nor destroyed; it can only be rearranged. Whether it is a Shakespearian play, a Beethoven symphony, or a hairy-nosed wombat, creativity often resides in the production of highly nonrandom systems.

Another common misconception is that evolution acts for the benefit of the species. Since natural selection operates by the differential reproduction of individuals whose characteristics render them more fit than others, the ultimate result of this process is a population composed of individuals each of whom is more fit than it would be if a nonselective (e.g., random) process were at work. In this sense, then, a species is better off because of evolution; inappropriate genetic combinations are weeded out and advantageous ones are encouraged. But this effect on the species as a whole is simply the cumulative result of natural selection operating independently on the individuals comprising the species. Evolution does not guarantee perpetual life to any species; in fact, the vast majority of species that have existed on earth are now extinct.*

On an intellectual level at least, human beings readily tend to accept the idea of subordinating individual interest to that of the group or species. This may in part be due to socialization processes that encourage such an attitude. In any case, there is little support for the idea that individuals will be selected for those behaviors that benefit the species. Whenever conflict occurs between benefits to the individual and benefits to the species, natural selection can most readily get a handle on the former. Benefits to a species, then, are largely incidental by-

---

* This is reminiscent of the observation by the satirist Tom Lehrer: "It is a sobering thought to realize that when Mozart was my age . . . he had been dead for six years!"

25

products of natural selection, which itself operates simply to maximize the reproductive success of each individual comprising that species. In this case, there is no "whole" other than the sum of its parts.

For example, male bighorn sheep have evolved behaviors and appropriate structures (horns) for spectacular head-butting (Plate 2.1). These extraordinary battles determine the dominant individuals who will then do most of the mating (Geist, 1971). Bighorn rams do *not* subject themselves to such annual headache rituals because of their concern that the best males be chosen so that the species will ultimately benefit. Rather, each individual engages in social competition because of the evolutionary payoff accruing to the winners: males who have been successful in such competition have produced more offspring than have unsuccessful males, so the genetic blue-print for such tendencies has spread and the species has come to consist of individuals that behave this way. Similarly, dominant males refrain from injuring smaller, weaker juvenile males. This may benefit the species by reducing mortality, but the basis for this restraint is almost certainly the ultimate reproductive disadvantage accruing to the individual who injures a close rela-

PLATE 2.1 TWO ADULT MALE BIGHORN SHEEP CLASHING HORNS.
(PHOTO BY V. GEIST)

tive and/or unnecessarily risks personal injury (see Chapter 4 for a detailed treatment of this important but difficult topic).

Let us take another example: young elephant seal pups are often squashed during violent fights between the huge adult males (LeBoeuf, 1974; Plate 2.2). This is a major source of mortality and it seems clearly disadvantageous to the species and even to the individual adult males, because the innocent victims are often their own offspring. However, such accidental infanticide occurs largely when males are defending their hard-won harems against potential usurpers. In such situations the potential loss of all his females constitutes a much greater detriment to each male's evolutionary success than does the occasional loss of an infant. So natural selection has favored egotistical concentration on preventing the former, with relative disregard of the latter. Of course, such a strategy would not in itself cause species extinction because, if it actually reduced the reproductive success of participating males, then individuals would be selected for showing somewhat greater regard for their offspring under foot or, rather, under flipper.

The role of natural selection in promoting evolutionary change may be further clarified by contrasting it with a major

PLATE 2.2 SOUTHERN ELEPHANT SEAL BULL ROLLING OVER A PUP.
(PHOTO BY B. LE BOEUF)

alternative view, Lamarckism. The Darwinian view of evolution is that genetic diversity is produced by mutation and recombination, following which the most successful products are selected naturally by leaving more offspring than their less successful counterparts. By contrast, Lamarckism posits that evolutionary change occurs via the "inheritance of acquired characteristics". Thus, in a classic example the ancestors of modern-day giraffes were imagined to have short necks that were stretched in attempting to reach high-growing leaves. Adult giraffes, during their lifetimes, thus acquired longer necks that in turn were inherited by their offspring. Similarly, if parents developed large muscles through exercise, this should contribute toward larger muscles in their future offspring. Despite an attractive plausibility to such notions, there is no evidence that changes in an individual's body can be passed on to his or her offspring. The flow of information during biological development is from the genes to the body and not vice versa.

Numerous studies have failed to confirm Lamarckian inheritance. For example, research in which the tails of successive generations of mice were amputated failed to produce any decrease in tail length among future offspring. (I might also add that centuries of circumcisions among Jews and Moslems have similarly failed to shorten the foreskin.) As with the Abominable Snowman or North American Bigfoot, research on Lamarckian inheritance can never disprove the *possibiltiy*; it can only evaluate specific attempts, and so far nothing has been confirmed. However, Lamarckism seems so logical that many people assume its occurrence without realizing it. Thus, we have the common assumption that the human species is evolving toward big-brained, small-bodied individuals, since we use our brains so much and our bodies so little! This presupposes heritable consequences of use and disuse, the inheritance of acquired characteristics. In fact, human evolution would proceed in this direction only if individuals with bigger brains and smaller bodies produced relatively more successful offspring, i.e., if directional selection of this sort was taking place. There is no evidence that it is.

Similarly, most people would attribute the common absence

of eyes among cave-dwelling fish, amphibians, and invertebrates to the simple fact that in their dark environment, these animals lost their eyes because they failed to use them. Lamarckism again. What almost certainly happened is that the sighted ancestors of modern day blind cave species were *selected against* once they began inhabiting pitch-dark environments, since eyes are delicate and susceptible to infection, disease, and injury. Of course, under normal lighted conditions, the selective advantages of eyesight greatly outweigh these disadvantages of eyes. But in a new environment selection progressively favored individuals that, because of their own mutations and recombinations, had eyes that were smaller and thus less of a hindrance to reproductive success. Of course, in normally lighted environments, such small-eyed individuals were regularly selected against, thus contributing to the relatively good vision we associate with most animals.

Despite the inadequacy of Lamarckism, it has enjoyed occasional scientific esteem, most notably during the time of T. D. Lysenko in the U. S. S. R. This was an extreme example of the politicization of science, as Lamarckism was more acceptable to Soviet authorities than Darwinism, presumably because of the former's egalitarian ethos (see Zirkle, 1949 for a fascinating account of this tragic era in Soviet genetics.) Lysenko attempted to breed winter-resistant grain by exposing seedlings to cold, allowing them to mature and to produce seed, then exposing the next generation of seedlings to cold again, and so on. This technique was not successful. By contrast, consider how plant breeders accomplish the same goal using the principles of natural selection: plants are exposed to cold, after which those individuals that show themselves to be most resistant are chosen (selected) and cross-fertilized among themselves, producing the next generation. It works. In fact, it is the same technique used naturally by any species to evolve a degree of resistance to any environmental feature: drought, heat, cold, predators, DDT, or penicillin.

The essential difference between the Lamarckian and the Darwinian approaches to genetic change is that the Lamarckian approach assumes that environmental experience will *change* the population's genetic makeup and it will do so in an appropriate, i.e., adaptive, manner. By contrast, the Darwinian approach does not seek to modify genotypes directly; the environment (cold, in the above example) is used as a selecting agent, which *reveals* the variability in genetic endowment already present in the population, due to mutation and recombination. Once identified, these "fittest" individuals are then selected to propagate the next generation, encouraging further concentration in their offspring of those genetic factors leading to the desired phenotype, in this case resistance to cold.

It is commonly believed that the human appendix is shrinking because we do not use it anymore. Clearly, this also is a Lamarckian interpretation, taken at face value. Yet something very much like this may be happening. We have already considered that living things constitute highly nonrandom systems produced and maintained by natural selection. In the absence of such "creative" forces, order tends to break down; this is the second law of thermodynamics. Any identifiable phenotype requires a nonrandom array of genes to specify the appropriate blueprint. This holds for a structure such as the appendix, and for behavior as well. The existence of such an organized array is assured by natural selection, but when environmental change removes the selective advantage of the phenotype in question, selection stops operating to maintain it. It falls into increasing disarray and can be expected to disappear eventually. Such nonadaptive phenotypes may therefore persist in a population for a time, but they will become increasingly vestigial, like the leg bones of a whale or the tendency of domestic dogs to circle before settling down on a living room rug, although the dogs no longer inhabit tall grasses occupied by biting insects.

Darwin and his contemporaries worked without any real knowledge of genetics. Then Mendel's basic laws were rediscovered about the turn of the twentieth century. Combined with identification of chromosomal structure and function and, finally, the elegant statistical formulations of population genetics (Fisher, 1958; Haldane, 1932; Wright, 1931; 1948), the modern *synthetic theory of evolution* was established (Huxley, 1942; Dobzhansky, 1951; Mayr, 1942).

One of the important cornerstones of population genetics and evolutionary theory is the Hardy-Weinberg Law, a simple statement of the relationship between gene frequencies and the frequencies of genotypes, the various possible combinations of these genes in different individuals. Basically, the relationship is that, if two alleles (alternative forms of the same gene) A and a are present in frequencies p and q, respectively, such that $p+q=1$, then the genotypes AA, Aa, and aa will occur in frequencies p^2, $2pq$, and q^2, respectively. Mathematically inclined readers will recognize this distribution as the binomial expansion $(p+q)^2$.

In the absence of factors that modify the frequency of genes, populations will retain this statistical distribution, i.e., they will be in Hardy-Weinberg equilibrium. Since evolution involves a *change* in gene frequencies, populations in such equilibrium are not evolving. From a population genetic viewpoint, therefore, the study of evolution is the study of those factors that disrupt Hardy-Weinberg equilibria. In brief, these factors are the influx and outflow of genes (immigration and emigration), mutation, assortative mating, statistical errors due to small sample size, and selection, the differential representation of certain alleles from one generation to the next.

Immigration obviously modifies gene frequencies by introducing new genes and/or altering the proportion of existing genes. Emigration has a similar effect by removing genes. Mutation has less obvious consequences; it also introduces new genes and eliminates old genes, but newly mutated forms also tend to mutate back to the original. Therefore, an increased frequency of a new gene produces an increased frequency of the original, ultimately see-sawing to a new equilibrium rather than producing significant evolutionary change in itself. Assortative mating is the tendency of like individuals to breed together. This will modify genotype frequencies by increasing the proportion of homozygous forms (AA and aa), but gene frequencies (A and a) remain unchanged. The two remaining factors are generally acknowledged to be the most important in producing evolutionary change: sampling error and selection.

Any statistical process is sensitive to the sample size being considered and genetics is no exception. Just as the results of flipping a coin will be less likely to approach equal numbers of heads and tails when only a few trials are made, a breeding population consisting of only a few individuals is less likely to represent all possible genotypes. Certain genes may even be lost. For example, if we flip a coin only twice we may get two heads, thus "losing" tails. Or vice versa. Similarly, a small breeding population may lose certain genes and underrepresent certain genotypes simply by chance alone. Gene frequencies under such conditions may thus drift randomly, nonadaptively, from one generation to the next: this is the phenomenon of *genetic drift* that has been implicated in apparently random, nonadaptive evolutionary changes possibly occurring in small populations (Wright, 1969).

The final and most important factor disrupting genetic equilibrium is natural selection, which we considered earlier in a different context. The Hardy-Weinberg Law assumes that each gene makes an equal contribution to the next generation, i.e., genes do not reproduce differentially. But, as discussed earlier, individuals differ in their reproductive success; so do genes. Alternatively, we could say that individuals differ in their reproductive success because of the differences in the genes they carry. Therefore natural selection acts to disrupt Hardy-Weinberg equilibrium, i.e., produce evolutionary change, in that certain genes produce phenotypes that are more successful than others and are better represented in succeeding generations. A measure of the selective quality of genes (or genotypes), their reproductive success, is their *fitness*.

In a stricter sense, fitness is a number that, when multiplied by the frequency of a gene or genotype in one generation, gives the gene frequency in the next. Thus, a fitness of zero means that individuals of that sort leave no offspring. A fitness of one means that the frequency in the next generation is unchanged from that in the preceding. Fitnesses of less than one indicate a decline in frequency, a production of offspring at less than the replacement rate, while fitnesses of greater than one indicate an increase in frequency, individuals reproducing at a rate that exceeds the rate that maintains the equilibrium.

For example, if an original population consists of 50 X individuals and 50 Y individuals and the next generation has 75 X's and 25 Y's, then the fitness of X is 75/50 or 1.5 and that of Y is 25/50 or 0.5 (50 × 1.5 = 75 and 50 × 0.5 = 25). Alternatively, evolutionary biologists often employ the concept of *relative fitness*, in which the highest fitness is defined as one, with the others being selected against to the extent that their relative fitness falls below one. In the above example, the relative fitness of X is 1.5/1.5 = 1 and that of Y is 0.5/1.5 = 0.33. We tend to think of fitness as an inherent characteristic of a particular trait, but in fact all fitnesses are relative, not only to other traits but also to the environment in question. A trait may confer high fitness in one environment but low fitness in another. For example, mutations for white fur increase the fitness of small rodents inhabiting White Sands National Monument; accordingly, many such animals are white in that environment. But in the nearby black lava beds a white coat is of low fitness, while black coats are reproductively more successful; hence small mammals living there have evolved black coats.

The concept of fitness is very important in evolutionary theory and in sociobiology, and it will be used extensively throughout this book. Another related notion is *adaptation*. An adaptation can be considered as any evolved characteristic of an organism that increases its fitness. Sociobiologists typically concern themselves with the *adaptive significance* of animal phenotypes, behavior in particular. This refers to the contribution of such traits to fitness, i.e., the nature and extent of evolution's impact upon the trait, and vice versa. A fundamental assumption of sociobiology is that behavior patterns are in fact adaptive. Indeed, this is the basis for our present concern with natural selection and the evolutionary process. However some debate currently exists among evolutionary biologists as to the extent of adaptive versus nonadaptive (random) factors in evolutionary change. Most opinion favors the former, and this book will be concerned with behavior as an evolved adaptation of living things. For a somewhat alternative view, see Lewontin (1974).

Finally, due to space considerations, I have limited this chapter to a cursory overview of the evolutionary process. For further information regarding the following areas, interested readers should consult additional sources: evolution in general, Grant (1963), Maynard Smith (1966), Moody (1970), and Stebbins (1971); genetic approaches, Ehrlich and Holm (1963) and Dobzhansky (1951, 1974); paleontology, Simpson (1949, 1953), Weller (1969), and Eaton (1970); speciation, Mayr (1963, 1970); mimicry, Wickler (1968); evolutionary ecology, Emlen (1972), Ricklefs (1973), and Pianka (1974).

THREE

Evolution and Behavior: Getting It Together

When a cow hears the buzz of a stable fly that bites, it merely ripples its skin or flicks its tail. By contrast, it reacts violently to the sound of a parasitic warble fly by prancing, kicking and jumping about excitedly. Yet biting flies inflict immediate and painful injury, whereas warble flies painlessly deposit tiny eggs. Why do cows show such great agitation in the latter case and so little in the former? If their behavior is the result of previous experience, why are they relatively more nonchalant about the painful biting flies? The answer is that the small, seemingly innocuous eggs of warble flies develop into larvae that will eventually burrow into the cow's tissues, causing serious disability and sometimes death. During their evolutionary history, cows that responded negatively to a warble fly's approach were less likely to be parasitized; therefore, they were reproductively more successful. Natural selection has thus promoted the evolution of a particular behavior in cattle, a behavior that is not otherwise explicable in terms of learning theory, socialization, or any of the other nonbiological approaches to behavior.

A rat, placed in a complex maze for the first time, will explore it although no food is present at one end. Later, when food is introduced, that rat will be able to learn the maze more rapidly than will a naive rat, one that is placed in the maze for the first time. Strange as it may seem, this finding has puzzled psychologists because the rat is not supposed to have learned anything the first time since there was no identifiable reward or *reinforcer*, such as the food on the second trial. For some inexplicable reason, the learning was latent in the initial, unreinforced experience. Puzzlement over the phenomenon of *latent learning* is an excellent example of scientists being captives of their own inadequate paradigm, in this case, reinforcement theory. By contrast, an evolutionary approach reveals that free-living rats normally occupy systems of runways and burrows: it is of great adaptive significance for such an animal to explore and familiarize itself with its surroundings, so that it can forage effectively and elude predators. Accordingly selection will have favored the inclination and ability to learn maze-like structures. Given this insight, it would be surprising if rats did *not* show latent learning of this sort.

Keller and Marian Breland, students of the psychologist B. F. Skinner, became professional animal trainers. They used the techniques of operant conditioning to induce raccoons, roosters, pigs, and other animals to perform curious and novel acts such as depositing coins in piggy-banks for a display in a bank window or playing a modified form of ping-pong. Occasionally however, the animals refused to cooperate: roosters made scratching movements with their feet, pigs snuffled incongruously on the bare floor, and the raccoons refused to deposit their coins, preferring instead to rub them between their paws in a manner most inappropriate to the inspiration of future savings account depositers. These animals had been trained by reinforcing them with food; however, when they were made hungry in an effort to eliminate the unwanted behaviors, the behaviors actually *increased*. The Brelands finally attributed these failures to the occasional surfacing of instinctive behaviors, adaptive aspects of each species' evolutionary past that could not be thwarted entirely, even by the sophisticated techniques of modern animal psychology (Breland and Breland, 1961). Thus, scratching with the feet, snuffling with the nose, and rubbing with the hands are evolved food-getting behavior patterns that are characteristic of chickens, pigs, and raccoons, respectively.

These examples all demonstrate cases in which an evolutionary perspective provides greater insight into the behavior of animals. It helps explain behaviors that fall outside the traditional paradigms of behavioral science. But its value is not limited to such peculiarities. Rather, the study of sociobiology employs evolution to interpret the basic patterns of animal social behavior and provides a rather sweeping synthesis of behavior, painting with broad strokes across a range of phenomena and species. An evolutionary approach involves a level of analysis different from that typically employed by social scientists. The two viewpoints can be distinguished conveniently as *proximate* causation versus *ultimate* causation.

In a sense, all students of behavior ask the question, "Why does an individual perform a particular behavior?" But there are different ways of answering this question. For example, we may ask why certain birds migrate south in the autumn and north in the spring. One way of responding would be to concern ourselves with the associated hormonal mechanisms. For example, a high sex hormone level is correlated with spring migration while the relative absence of such hormones correlates with autumn migration. In addition, the accumulation of body fat during late summer may be responsible for the migratory restlessness observed in the autumn. Alternatively, we may consider the environmental cues to which birds are sensitive: temperature, rainfall, and barometric pressure have some value as predictors of season, but amount of day light is more reliable. It is therefore interesting that a particular region of the brain, the hypothalamus, contains receptors that are sensitive to changes in day length and in turn influence the birds' physiology to prepare for migration.

These are all proximate mechanisms, explanations of behavior in terms of its immediate causation. We could also inquire legitimately into ultimate mechanisms: an evolutionary approach. With regard to migration, we might address ourselves to the adaptive significance of this behavior. In seeking to discover why birds migrate in this manner, we would inquire into the consequences of migration versus nonmigration insofar as the reproductive performance (fitness) of individuals is concerned. Migration is a dangerous and energetically expensive procedure. On the other hand, year-round occupancy of a particular breeding area also imposes some costs. An ultimate analysis of bird migration might involve such considerations as relative food availability at the breeding and wintering areas, climatic conditions, avoidance of predators and competitors, the added difficulty of reestablishing territorial ownership after seasonal absence, and the hazards of migration itself. Also, we might attempt to determine why the proximate mechanisms, suggested above, operate in certain animals, e.g., warblers, but not in others, e.g., chickadees.

Proximate analysis "explains" the hormonal, neural and/or stimulus-specific factors operating within an animal to produce a particular behavior. Ultimate analysis explains why these proximate mechanisms occur in the first place and/or why animals respond to them as they do, regardless of what they are. Proximate questions about behavior may include: How does behavior develop within an individual? In what way can experimental manipulations alter behavior? What stimuli elicit which behaviors? What are the genetic, physiological, and/or anatomical factors that influence behavior and how do they operate?

Likewise, ultimate questions about behavior include: What is the evolutionary history of a behavior in a population? What is the adaptive significance of the behavior? Are animals behaving in a way that maximizes their fitness? Why do similar (or different) species perform similar (or different) behaviors, and why is the behavioral repertoire of a particular species (plus age and sex class within each species) what it is and not something else? In fact, proximate mechanisms can profitably be viewed as the servants or tools of ultimate causes, as means to an end. Regardless of a researcher's interest, however, before either of these approaches can validly be attempted, it is necessary to describe accurately the normal behaviors of each species. Ethologists refer to such characterizations as *ethograms*. Similar descriptions of the unaltered social behavior patterns of animals may be called *sociograms* (Wilson, 1975).

In the long run, of course, explanations of behavior will be most satisfying when they take both proximate and ultimate causes into consideration. In the short run, it is a worthwhile exercise to step back periodically from any explanation of behavior and ask whether it is proximate or ultimate. Neither one is inherently better, but neither one alone is complete.

Sociobiology is concerned largely with interpreting behavior in ultimate terms. As discussed in Chapter 2, evolution can influence a trait only if there exists some correlation between the genotype and the phenotype in question, i.e., genes must somehow relate to behavior. This seems to be a difficult area for students to understand, especially those in the social sciences, so it will receive special attention here.

One of the oldest and least productive debates in the history of science concerned the underlying causes of behavior. A basic dichotomy of approaches can be identified, with proponents favoring either genetically mediated factors or those characteristics acquired during the lifetime of an individual. The conflict has been variously described as instinct versus learning, nature versus nurture, or endogenous versus exogenous control of behavior. Biologists, particularly ethologists, and social scientists, particularly psychologists, engaged in rather acrimonious debate, with the biologists favoring genetic influences and the social scientists emphasizing the role of learning, culture, and other environmental modifiers of behavior. Finally a compromise synthesis has emerged (see particularly, Hinde, 1970; Lehrman, 1970). It is now generally recognized that, like most dichotomies, the either-or question was essentially meaningless. Thus, it makes no sense to consider an animal's development and behavior in the absence of an environment; likewise, the extreme of environmentalists' claims would posit an environment without any organism. In short, all phenotypes derive from the interaction of an organism's genetic potential with its environment, and behavior is as good a phenotype as any other. This model is presented as a general one for all living things:

GENOTYPE

ENVIRONMENT

PHENOTYPE
(behavior, anatomy, physiology, etc.)

The relative contributions of genotype and environment may vary considerably, but neither is ever equal to zero.

For example, young male white-crowned sparrows learn the song that is characteristic of their species by hearing the singing of their fathers. If the nestlings are reared in acoustic isolation, they produce abnormal songs; it may therefore appear that the behavior is determined by learning alone. However, if an isolated bird is played recordings of a variety of sounds, including the normal white-crowned sparrow song, it will then sing normally itself. Without any previous experience in this regard, the young bird chooses the song that is appropriate for its species and selectively learns that one rather than another. It is as though the bird has some sort of internal template that permits it to recognize the appropriate behavior when it is experienced, even though some minimal experience is necessary in order for the normal behavior to occur. Behavioral phenotypes differ with regard to their particular pattern of interaction between experience and genotype, but some interaction occurs in every case.

As a simple graphic model, assume that the total accumulated influence of genotype and environment must be a constant in order for a behavior to occur. It may be conceptually clearer to imagine a genotype and an environment as each specifying information to an organism, with each behavior requiring a certain amount of information so that it occurs. Generalizing somewhat crudely across species, we could array each on a graph showing the relative contributions of the two primary factors to the behaviors of each species (FIG. 3.1).

The simplest unicellular organisms would obviously be at the extreme of maximum genetic control, minimum flexibility, and humans would occupy the other extreme. The relative locations of various other species could be estimated; e.g., mammals would be closer to humans, fish closer to invertebrates, etc., although none could be placed exactly. And for no species would either the environmental or the genotypic component equal zero. We could also examine the behavioral repertoire of a single species: a similar diversity is generally present. Among humans, for example, the blink reflex is highly specified by our genetic makeup whereas personality is largely determined by experience. But, again, the influence of genotype probably never reaches zero, not even for the most complex human behavior (see Chapter 10 for a more detailed consideration of this controversial topic).

Another way of viewing the interaction of genetics and environment in producing behavior is to recognize that behavior is not contained somehow within a gene, waiting to leap out like Athena, fully armored, from the head of Zeus. Rather, genes are blueprints, codes for a range of potential phenotypes. In some cases the specification may be very precise, leaving little room for modification due to learning or other experiences. In others, the blueprint may be so general as to be almost entirely at the disposal of experience. Nonetheless some restrictions remain; an armadillo can behave only like an armadillo and a

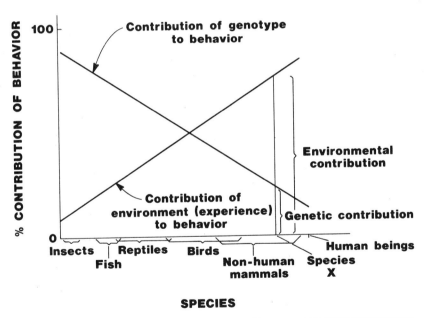

FIGURE 3.1. RELATIVE CONTRIBUTIONS OF GENETIC AND ENVIRONMENTAL FACTORS TO THE BEHAVIOR OF DIFFERENT ANIMALS. For any given species (X), the sum of genetic and environmental components is 100%. (Technically, a third component should be added: an interaction effect between genotype and environment.) Note that genetic and environmental contributions are each above zero for every species.

41

zebra must behave like a zebra. To some extent, we can identify certain characteristics of an animal that predispose it to relatively more or less reliance upon genetic specification. For example, tight specification of the route from genotype to phenotype would be expected among organisms having simple nervous systems, little capacity for complex movements, and that experience relatively simple and unchanging environments. By contrast, the ability to modify behavior as a consequence of previous experience, learning in a broad sense, requires possession of a relatively complex nervous system. The need to interact with a variety of environments and the physical capacity to do so in complex ways also select for flexibility, that is, susceptibility to modification by experience.

Within a single species, the optimum degree of genetic specification varies depending on the behavior in question. For example, behaviors that must be performed correctly the first time, such as escape from predators, are likely to be under somewhat precise genetic control. There is generally little opportunity for prey species to rely on experience to refine progressively their anti-predator responses. Individuals that possess relatively flexible blueprints for anti-predator behaviors would accordingly be less fit than those relying on a more canalized, genetically-specified system. In contrast, among species such as wolves, in which the preferred food (e.g., moose) is so large that the young cannot obtain it for themselves but must rely instead upon parental provisioning, there is little room for rigid specification of food preferences among the young. Further, a wolf pup need not possess a precise, genetically programmed recognition of its food or how to obtain it. In fact, it might be less fit if it were too finicky, because available prey may vary depending on environmental conditions each year. Furthermore, because of the prolonged dependency of wolf pups combined with the cohesive social organization of each pack, a young wolf has great opportunity to modify its behavior via its experience. In contrast, after hatching from its egg, a young spider is likely to float away on the first good breeze and will have to function independently. Spiders must therefore possess fully developed behavioral blueprints; reliance on instincts is more adaptive for them than it is for wolves, although wolves certainly possess some instincts as well. Similarly, humans rely even less on instincts. In a sense, then, we can interpret the various patterns of behavioral specification as representing different evolved strategies, each of which produces the maximum individual fitness in each case.

Given that behaviors differ in their reliance on genetic control, we would expect them to differ in their responsiveness to selection, with a positive correlation between such control and the ability of selection to alter the trait in question. In other words, the more tightly a phenotype is bound to a genotype, the greater the anticipated impact of differential reproduction on the distribution of that phenotype in succeeding generations. Whenever selection occurs, the magnitude of the resulting change is determined primarily by two factors: the difference between the mean of the original population and those individuals selected as parents, and the responsiveness of the particular trait to genotypic changes. If the selected parents differ from the population mean by S, the *selection differential*, and the offspring of these parents differ from the original population by R, the *response to selection*, then R/S is a measure of the effect of selection in altering the trait (FIG. 3.2).

The trait is very responsive to selection if a small selection differential (S) produces a relatively large change (R) in the distribution of the trait, i.e., if R/S is large; however, R/S cannot exceed one. Alternatively, if a large selection differential produces only a small response, then the trait is clearly not very responsive to selection. The value of R/S can also be taken as a measure of *heritability*, the proportion of phenotypic variance attributable to genotypic variance. As such it is a characteristic of populations, not of individuals. The responsiveness to selection as a function of selection differential provides a measure of heritability. Alternatively, the fact of heritability is what permits selection to be effective in generating behavioral change. If $R = S$ (the new generation exactly resembles the mean of the selected sample), then heritability is 1 and all phenotypic variance is attributable to genotypic variance. On the other hand, if $R = 0$, then selection has been ineffective. In such a case, the

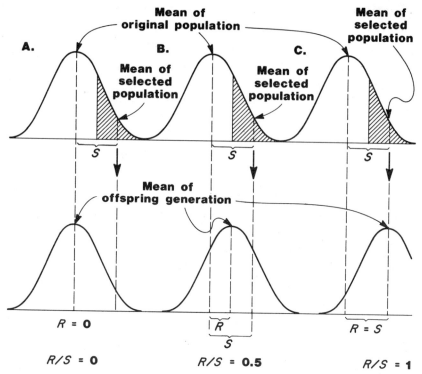

FIGURE 3.2. THE RELATIONSHIP BETWEEN SELECTION DIFFERENTIAL (S), RESPONSE TO SELECTION (R), AND REALIZED HERITABILITY (R/S). (A) The mean of the offspring generation does not differ from that of the original population; $R = 0$. Therefore, $R/S = 0$. (B) The mean of the offspring generation differs from the mean of the original population by an amount R, where R is less than S. In this case, $R/S = 0.5$. (C) The mean of the offspring generation equals the mean of the selected population. In other words, $R = S$, so $R/S = 1$.

distribution of the trait in the original population must have been independent of the distribution of genotypes within that population, i.e., it must have been due entirely to environmental effects acting somewhat differently upon different individuals. Heritability would therefore be zero. For essentially all phenotypes, however, including behavior, heritability tends to be between zero and one.

There are several ways of studying the correlations between genetics and behavior, ranging from the effects of single genes and mutations (Rothenbuhler, 1964; Ehrman and Parsons, 1976) through analysis of embryonic development (Benzer, 1971, 1973) to actual identification of chromosomes associated with particular behaviors (Hirsch and Erlenmeyer-Kimling, 1962). With regard to complex behaviors, perhaps the most useful techniques have been artificial selection and hybridization studies. Artificial selection for a behavior trait simply involves directional selection, with humans serving as the selecting agent; it has been a standard procedure among animal breeders for hundreds of years. In the laboratory, artificial selection has been conducted for the tendency to move toward or away from gravity in fruit flies (Hirsch, 1963), aggressiveness and mating behavior in domestic fowl (Siegel, 1972; Cook et al., 1972), various degrees of alcohol preference in mice (Lindzey et al., 1971), courtship and mating speed in fruit flies (Manning, 1965), emotionality and tendencies for exploratory behavior in mice (Lindzey and Thiessen, 1970), and a remarkable range of behavior traits in domestic dogs, cats, horses, etc. (FIG. 3.3). In fact, all studies conducted thus far attempting to select for any behavior trait among animals have been successful in changing the distribution of that trait in the experimental population. The implication is clear: what artificial selection can do, natural selection can also do. Evolution in nature is likely to be slower than in the laboratory, because artificial selection can concentrate upon a small number of desirable traits, whereas among free-living animals extreme development of one characteristic may be prevented by constraints upon the whole animal to maintain its overall fitness. But the essential factor of differential reproduction remains unchanged.

Hybridization studies have also revealed undeniable genetic influences on behavior. In such studies, parents are chosen from different species or different varieties within a single species; all that is necessary is that they represent genetically distinct populations that differ with regard to certain behaviors. The hybrid offspring produced by such cross-breeding frequently behave similarly to one parent or the other, or in some cases they are intermediate between the two. The precise pattern depends upon the nature of the genetic influence, whether

45

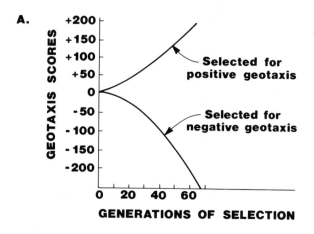

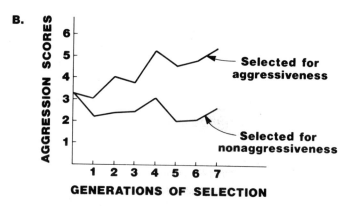

FIGURE 3.3. RESULTS OF SOME SELECTION EXPERIMENTS FOR BEHAVIOR. (A) Selection in fruit flies for positive versus negative geotaxis, the tendency to go upward or downward. (*Modified slightly from Erlenmeyer-Kimling et al., 1962.*) (B) Selection in mice for aggressiveness versus nonaggressiveness. (*Modified slightly from Lagerspetz, 1964.*)

a single gene or many genes, discrete or continuous effects, etc. The important point for our purposes is that any consistent pattern suggests the role of genes in mediating the behaviors in question. Successful studies of this sort have included courtship behaviors in fish (Clark et al., 1954) and ducks (Sharpe and Johnsgard, 1965), vocalizations in crickets (Bentley and Hoy, 1972), and an enormous diversity of behaviors in laboratory mice and rats.

Genetic influences upon behavior were dramatically apparent in a study of nesting behavior of some African parrots, commonly known as lovebirds (Dilger, 1962). Two species of these parrots differ in the manner in which they carry material used in constructing their nests. Members of one species carry pieces of grass, paper, and other nesting material in their beaks, while members of the other tuck such things in their rump feathers.

Hybrids with one parent from each species have been produced. When it comes to carrying nest material, these offspring are curiously befuddled. They attempt an inappropriate combination of both parental techniques and are unable to do a good job of either. Thus, they pick up materials and point toward their rump feathers, only to drop them. Or they insert the nesting materials correctly, only to remove them immediately afterwards. This kind of intermediate behavior pattern among genetically intermediate individuals provides strong evidence for the role of genotype in mediating behavior. It also suggests one reason why hybrids are rare in nature: Hybridization is doubtless selected against (naturally) in many cases because the genetically intermediate offspring are behaviorally intermediate and, consequently, inadequate.

In summary, to many persons the term "behavior genetics" may seem to be a *non sequitur*, but it is a vigorous area of research (see Ehrman and Parsons, 1976, for a comprehensive introduction). If the correlation between genes and behavior still appears perplexing, despite the overwhelming evidence, consider it from a purely mechanistic viewpoint: the DNA of which genes are composed specifies the production of proteins, leading to the various structures constituting an organism. These structures include bone, muscle, gland, and nerve cells. Behavior unquestionably arises as a consequence of the activity of nerve cells, which presumably are susceptible to specification by DNA, just as any other cells. Accordingly, insofar as genes specify the organization of nerve cells, just as they specify the organization of bone cells, there is every reason to accept a role of genes in producing behavior, just as we accept a role of genes in producing structure. As phenotypes go, behavior may be somewhat more flexible or susceptible to environmental influences than most. But the relevance of genetics to behavior is undeniable, and since evolution is the primary force responsible for the genetic make-up of living things, evolution must also be relevant to behavior.

Therefore, behavior patterns are subject to evolution by natural selection, just as any other phenotype. Differential reproduction of individuals demonstrating characteristic intensity, amplitude, frequency, or threshold for a particular trait is very likely to modify the distribution of that trait among individuals comprising the next generation. At first glance, certain complex instinctive behaviors may appear too elaborate to have evolved through natural selection. However, an eye or a brain is equally complex and in fact, the phenomenon of *genetic assimilation* (Waddington, 1960) provides one of several possible mechanisms for the evolution of complex instinctive behaviors. Thus, the minimum genetic specification of a behavior would involve simply *permitting* it to occur, given the appropriate environmental conditions. Let us start with natural selection of individuals, each of which initially is capable of behaving in certain ways at certain times, e.g., learning to make a difficult discrimination. Differential repoduction of such individuals combined with sexual recombination among their offspring could soon generate individuals who possess genotypes that are increasingly relevant to the behavior in question, e.g., permitting it to occur at a progressively lower threshold until eventually individuals are produced whose genotypes encourage them to perform that behavior even without the experience that their ancestors required; the behavior is now an instinct.

Although learning is often considered antithetical to instinct, the two may be closely related in that tendencies to learn are themselves influenced by genotype. For example, the psychologists, Tolman (1924) and Tryon conducted now classic studies of artificial selection for maze-learning ability in laboratory rats, eventually deriving two distinct strains: maze-bright and maze-dull (FIG. 3.4). Furthermore, these two strains were maintained separately and their descendants were tested again in the original maze, many generations later. They still retained their

48

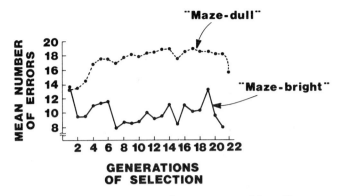

FIGURE 3.4. THE RESULTS OF SELECTION FOR MAZE-BRIGHT AND MAZE-DULL RATS. (*Modified slightly from McClearn, 1963.*)

bright and dull performances, respectively. But when tested in a maze that was slightly different from Tryon's original one, there was no significant difference between the two populations (Searle, 1949)! The implications of this finding are obscure, but it suggests the caution that selection for behavior may actually involve finer and more precise distinctions than we, as selectors, are aware. Thus, the initial study may have been selecting originally for a specialized tendency rather than some generalized brightness or dullness itself. For example, if the original maze required that animals make two right turns followed by a left, then brightness in learning this maze may have consisted simply of the tendency to do that, as opposed to intelligence, which should generalize to other mazes as well. Selection can indeed change behavioral phenotypes, but it remains unclear whether such changes are incidental by-products of other, more specific factors on which selection is operating more directly.

In any case, behavior can evolve, just like any other characteristic of a species. Some of my own observations suggest an instructive example of selection apparently causing an evolutionary change in behavior within a free-living population of animals. Most species of the common fruit fly *Drosophila* characteristically engage in elaborate courtship rituals prior to mating. Typically the male approaches the female and initiates courtship by vibrating his wings for varying lengths of time. On certain small, windswept islands off the coast of South America, males seem to engage in their form of courtship for significantly less time than do males inhabiting the mainland. The exact basis for this difference is unknown, but a comparison of the mainland and island populations suggests that directional selection has taken place in deriving one distribution from the other.

Thus, let us assume that the ancestral fly population inhabited the mainland. A small sample of these individuals, taken at random, colonized an offshore island; perhaps they were blown there by a tropical storm. Small coastal islands often lack trees and other large vegetation; on the average, they experience much stronger winds than does the mainland. Therefore, it is likely that within the new island population, males that courted too long were literally blown away, thereby selected against relative to the more fit males that demonstrated briefer courtship. Therefore, selection favored brevity of courtship so that an evolutionary change occurred, eventually producing a population with a different distribution of the behavior trait in question. This explanation is admittedly speculative but it is in accord with the observed facts. In addition, abundant evidence exists for a genetic influence upon courtship in these animals (Manning, 1961), and significantly, the insect fauna of small, windswept islands commonly contains a large proportion of wingless forms.

In addition to suggesting natural selection for a behavior trait among free-living animals, the above example is also instructive in that it suggests an important generalization of evolutionary biology: the need for adaptive compromise. Thus, we might ask why the evolved decrease in courtship duration of island dwellers stopped where it did. Why did courtship duration not become shorter yet? And why did it become as short as it did? Just as in other aspects of life, in evolutionary biology matters are rarely black or white; there are advantages and disadvantages to every trait, measured ultimately in terms of fitness, and we can consider that selection for each trait will maximize the difference between advantage and disadvantage. In our *Drosophila*

50

example, one advantage of brief courtship is a reduced likelihood of being blown away; a disadvantage is a reduced likelihood of stimulating a female sufficiently to achieve mating. Presumably then, the need for a certain amount of this behavior establishes limits to courtship brevity in the island population. Similar evolutionary pressures may well have been operating within the mainland population as well but with different intensities, because wind was a less significant factor. Accordingly, the distribution of courtship duration differed adaptively in the two populations. Stabilizing selection eventually operated on both populations, but the resulting distributions differed as a function of differences in the two environments.

Evolutionary analyses of behavior are rarely as clear-cut as those described so far. For a hint of the potential complexity, consider the selective pressures operating upon females with regard to their preference for males having differing courtship durations. On one level, females should not concern themselves with the fate of their consorts, as their only interest is to ensure fertilization of their eggs. However, females would enhance their fitness if they produced offspring who were themselves more fit, i.e., if their offspring possessed characteristics that enabled them to produce more offspring in turn. Given that short-duration male courtship is an adaptive characteristic in the island environment and that this trait has a substantial heritability, male offspring of quick-courting males should themselves tend toward quick courtship and therefore be fitter than their long-courting counterparts. Therefore, island females should be selected for preferring short-courting males because their male offspring would accordingly tend toward quick (adaptive) courtship, thereby providing these initial females with enhanced fitness via their *grandchildren*, i.e., the offspring of their sons rather than their offspring directly.

It should be emphasized here and remembered throughout all evolutionary analyses of behavior that no assumptions need be made about the internal motivational state of the individuals concerned. Thus, it is a convenient shorthand to use such expressions as "concerned with", "has an interest in", "is better off by doing", or even "wants to". These expressions do *not* imply cognition or volition. They are simply less clumsy than saying, "has been selected for responding in such a manner because ancestors behaving this way were more successful reproductively than were those that behaved differently". Organisms doubtless are selected for behaving *as though* they are aware of how to maximize their fitness, but this does not require any assumptions as to what is going on inside their heads. It is expected that they do what is appropriate; it is neither required nor expected that they know why.

Some insight is gained by acknowledging that evolution is a major factor that guides behavior. However, to be valuable, a paradigm should do more: it should organize present thinking and also suggest further insights. We will now explore four basic ways in which the study of behavior can make use of evolutionary theory. The following is the first attempted categorization of these approaches: (1) *historical*, (2) *evaluative*, (3) *correlational*, and (4) *predictive*. These endeavors are not necessarilly discrete, and in fact there is much conceptual overlap, particularly among (2), (3), and (4).

The historical approach involves the identification of presumed behavioral phylogenies, a reconstruction of the evolutionary history of the behavior in question. Such efforts face enormous difficulty because animal behavior leaves few fossils (for fascinating exceptions, see Seilacher, 1967). Therefore, the historical approach must employ comparisons of behavior in closely related extant species and must rely upon other criteria to provide insight into the presumed ancestral sequences. Two commonly used criteria are simplicity and commonality. All things being equal, biologists assume that simpler behavior patterns are ancestral to more complex ones. Similarly, behaviors that are found among many closely related species are generally assumed to be relatively primitive as opposed to behaviors unique to a small number of species. The latter are considered to be secondarily derived and are often interpreted as unique adaptations, specialized to meet particular environmental conditions.

For an especially delightful and perhaps instructive example of the historical approach, consider the courtship displays of the empidid flies (Kessel, 1955). In several species of this family of small, predatory flies, the male initiates courtship by presenting the female with an elaborate silken balloon of his own making. Why? Proximate answers might involve the role of prior experience, the neural and hormonal control of this behavior, or the particular stimuli to which males are sensitive. The ultimate causes for this peculiar behavior are entirely obscure until we look at other empidid species. In some of these, the males simply approach the females and begin courtship in ways similar to *Drosophila*, described above. But most empidids are predatory, with the females larger than the males, and not uncommonly an unlucky male winds up being a meal rather than a mate. The males of certain species avoid this fate by obtaining prey and presenting their chosen female with this morsel; while she feeds on it, he is able to copulate unmolested (Plate 3.1). Among other species, the prey is first adorned with silk, perhaps making it more conspicuous to the female. In others, the prey is surrounded with a silken balloon; this may ensure that the females take longer to open the "gift", giving the males more time to complete copulation. Given the effectiveness of the elaborate wrappings, the males of certain species not surprisingly first suck the prey item dry before presenting it to their paramours (Once the "deception" is discovered, it is too late!). Finally, others dispense with the prey altogether, simply spinning an elaborate silken balloon with which the males go a'courting. Astute readers will recognize this as our starting point. They will also recognize that without the historical interpretation provided by a comparison of closely related species, this behavior would remain a mystery.

Behavioral phylogenies have been discussed for numerous species, including Fiddler crabs (Crane, 1943), wasps (Evans, 1962), Galapagos finches (Lack, 1947), and ducks (Lorenz, 1958). However, the historical approach has its weaknesses. In most cases, the proposed phylogenetic sequences seem eminently reasonable, but in the absence of fossil behavior patterns there is no definite way to refute or verify the assumption of which behavior came first.

PLATE 3.1 FEMALE SCORPION FLY ABOUT TO FEED ON A PREY ITEM JUST GIVEN HER BY THE MALE. Although these animals are not empidids, they share the same behavior—males present females with captured prey prior to copulation. (PHOTO BY R. THORNHILL)

The evaluative approach differs from the historical in that it is more explicitly functionalist. It attempts to evaluate the adaptive significance of observed behavior patterns. In many cases this is easier said than done, and there is a danger in being too easily convinced by one's own cogent-sounding speculations, which nonetheless remain unverified. One of the better examples of the evaluative approach comes from work of the pioneering ethologist Niko Tinbergen (1963). He observed that adult black-headed gulls invariably removed empty egg shells that were left after their young had hatched, and they always did so within a few minutes after hatching.

This sanitation rite, although seemingly trivial in itself, was stereotyped, invariant, and highly predictable. Furthermore, it exposed the young to possible predation while the adult was away. The great species-typical consistency of egg shell removal suggested a strong genetic basis, and the risk suggested that the behavior itself must be of substantial selective importance. Therefore Tinbergen decided to evaluate the adaptive significance of egg shell removal. Having eliminated several other possible explanations, he decided that a reasonable ultimate interpretation of this behavior is that it helped protect the chicks by rendering them less conspicuous to predators. Thus, gull eggs are typically well-camouflaged with a mottled pattern, although the inside of the shell is white, making it easy to see. Removal of these conspicuous objects after hatching might accordingly serve to maintain the highly evolved camouflage.

By establishing experimental nests with empty, broken shells at different distances from the intact eggs, Tinbergen demonstrated that eggs in close proximity to such shells were significantly more likely to be detected by predators than were nearby eggs with intact shells. It therefore seems reasonable that parent gulls who remove shells run a smaller risk of losing their offspring to predators than do those who refrain from such behavior. This interpretation is further enhanced by the observation that kittiwakes (gulls specialized for nesting on cliff edges where predators are rare) do not engage in egg shell removal.

For another example of the evaluative technique in sociobiology, we might ask why many small birds are monogamous, the basic reproductive unit consisting of one male and one female. Because of the high metabolic rate and consequent food needs of developing nestlings, the cooperation of two parents may well enhance the survival and ultimate success of the young, which in turn increases the fitness of the parents. Monogamy may be adaptive for these species, because it ensures a greater food supply for the young. This is supported by the observations that, among monogamously mated species, generally both parents share in provisioning their offspring and that a single parent is often unable to provide as much food as a mated pair (Martin, 1974), resulting in lower reproductive success. This evaluation is further enhanced by the observations that among most polygamous and promiscuous birds, species in which one parent typically cares for the offspring, food is either present in great abundance, e.g., marsh-nesting species such as red-winged blackbirds, so that a single parent can forage adequately for its brood, or the young are relatively precocious at hatching and can forage for themselves, e.g., pheasants and grouse. Of course, numerous other factors may be involved (see chapter 6 for a more detailed treatment of the adaptive significance of animal mating systems), but evaluative considerations of this sort nonetheless permit significant insight into behavior (Plate 3.2).

In this example, confidence in our evaluation of the adaptive significance of avian monogamy was enhanced by comparison with mating systems in other species. Similarly, Tinbergen's evaluation of egg shell removal in gulls was supported by the absence of such removal in cliff-nesting species. Comparisons between species are thus important additions to the evaluative approach. This suggests a link to the third general way of using evolution to study behavior, the correlational approach.

PLATE 3.2 MALE PIED HORNBILL FEEDING INCUBATING FEMALE. In this monogamous species, male and female cooperate in walling off the female inside a tree cavity, within which she lays and incubates her eggs. This arrangement protects her and her offspring from predators, especially snakes, while they are most vulnerable. The male makes an essential contribution to the pair's ultimate reproductive success, in that he is the sole of source of food during this time. The tip of the female's bill can barely be seen protruding from the lower left edge of the narrow nest opening. (PHOTO BY B. NIST)

Insofar as behavior patterns are subject to evolution, they should be adaptive, i.e., given the available alternatives, they should confer maximum reproductive success upon their practitioners within the environments in which they evolved. There should also be discernible correlations between behaviors and certain environmental parameters, if we compare species that are closely related and that occupy environments that differ in some consistent, identifiable ways. By choosing closely related species, different species within the same genus or perhaps different subspecies within the same species, we minimize the idiosyncratic uniqueness of each behavior-environment match-up. Analysis can then concentrate on the presumed influence of different environments in selecting for different behavioral responses. This is the heart of the correlational approach, correlating behavioral differences among closely related populations with differences in their environments.

One example of this approach comes from studies of the social behavior of marmots (Barash, 1974a). These medium-sized, ground-dwelling rodents lend themselves well to sociobiological studies. They are large enough to be easily observed; they are diurnal (active in daylight); and they are relatively insensitive to the presence of a human observer. This trait minimizes a serious problem in sociobiology, a sort of Heisenberg's uncertainty principle applied to behavior, wherein the presence of the observer introduces possible changes in the behaviors being observed. Most important for our purposes, however, is that marmots include several closely related species inhabiting a variety of environments. As expected, there are consistent correlations between behavior and environment in each case.

Woodchucks inhabit low-elevation fields primarily in the eastern United States, and they are relatively aggressive and solitary animals. A male and female associate only briefly, at mating. The basic social unit consists of one adult female and her young, and even the young disperse from their natal burrow when weaned. Adults are socially intolerant of each other. In contrast, Olympic marmots inhabit high-elevation alpine meadows in Olympic National Park. In this stressful, short growing season environment, Olympic marmots are remarkably nonagressive. They live in distinct, well-integrated colonies and are socially tolerant (Plate 3.3). Olympic marmot social behavior is characterized by play-fights, greetings, and a relaxation of dominance and territorial exclusiveness (Barash, 1973a). The social intolerance of low-elevation woodchucks and the relative tolerance of high-elevation Olympic marmots suggest a possible correlation, one that may well be adaptive. This presumptive correlation is confirmed by the observation that a third species, yellow-bellied marmots, which inhabit intermediate-elevation environments, are correspondingly intermediate in their social behavior. These animals are colonial but nonetheless somewhat intolerant (Armitage, 1962; 1974).

It is one thing to recognize a correlation; the next step is to determine whether natural selection can be identified as its

PLATE 3.3 A GROUP OF OLYMPIC MARMOTS INHABITING AN ALPINE MEADOW. From the left, they consist of an adult male, two adult females and one juvenile. Groupings of this sort are common among this species, whereas they are not found among woodchucks inhabiting low elevation environments. (PHOTO BY D. BARASH)

58

cause, i.e., does the correlation have adaptive significance? The adaptive nature of the behavior-environment correlation among marmots is suggested by other correlations: Woodchucks (low elevation) become sexually mature as yearlings, and they generally disperse within their first year; yellow-bellied marmots (medium-elevation) mature during their third summer and disperse during their second; and Olympic marmots (high-elevation) mature during their fourth summer and disperse during their third. In addition, as we go from woodchucks to yellow-bellied marmots to Olympic marmots we find a progression in the age required to achieve adult weight. Table 3.1 summarizes the essential correlations thus far identified for these animals.

The most important environmental factor appears to be not elevation itself but, rather, the amount of food available to each species during a year. There is good evidence that aggression from adult marmots is instrumental in causing the young to disperse, to leave their colony of birth. It would clearly be adaptive for the adults to refrain from any aggressiveness that would precipitate the suicidal dispersal of their own undersized young. The progressively increasing need for delayed dispersal among marmot inhabitants of environments with progressively less available food may at least in part explain the striking correlation of marmot social systems with their environments.

Correlational studies of this sort are making useful contributions to our understanding of the impact of evolution upon complex patterns of animal social behavior. Analyses of environment-behavior correlations within closely related species have been conducted for primates (Crook and Gartlan, 1966; Eisenberg et al., 1972), mountain sheep (Geist, 1971), antelopes (Jarman, 1974), blackbirds (Orians, 1961), Arctic sandpipers (Pitelka et al., 1974), and numerous other animal groups. Given the existence of these correlations, we can expect such studies to be pursued vigorously in the future. But as with the historical and evaluative approaches, the correlational approach is often difficult to verify. This is not the fault of the researchers; it is inherent in the questions asked and in the nature of the biological world under study.

TABLE 3.1

CORRELATIONS BETWEEN THE SOCIOBIOLOGY OF MARMOTS AND THE ENVIRONMENTS IN WHICH THEY LIVE

	ENVIRONMENT			SOCIOBIOLOGY			
SPECIES	ELEVATION	GROWING SEASON	AGE AT DISPERSAL	AGE AT SEXUAL MATURATION	REPRODUCTION	SOCIAL SYSTEM
Woodchucks	Low	Long	First year	Second year	Annual	Solitary, aggressive
Yellow-bellied marmots	Medium	Intermediate	Second year	Third year	Occasionally skip a year	Colonial, moderately aggressive
Olympic marmots	High	Short	Third year	Fourth year	Biennial	Colonial, highly social, tolerant

60

The correlational approach poses a subtle danger in that it reflects the mental adroitness of the investigators as much as it does natural selection. Thus, the presence of a consistent correlation suggests that *something* is going on; but, given almost any such correlation, a competent evolutionary biologist can generally point out how it is, in fact, adaptive. This problem is overcome in part by attempting to apply the correlational approach in a predictive manner. For example, consider some environmental parameter, such as abundance of food, predator pressure, or severity of competition for breeding sites, that varies regularly, perhaps as E, $2E$, $3E$, and so on, across an array of habitats. Furthermore, assume that for a particular array of populations in these different habitats, we have identified a correlated series of behavior patterns, such as number of females per male, average group size while foraging, average size of territory, intensity of intra-species aggression, etc, that varies as $10B$, $9B$, $8B$, and so on. Most likely, we will initially have identified correlations at relative end points of the environment-behavior scale, e.g., E and $8B$, $8E$ and B. Our correlational approach will be somewhat supported if an E between 1 and 8 is subsequently found to be associated with a B between 8 and 1. It might seem especially powerful if the prediction is quite precise, e.g., $4E$ associated with $5B$, but this is not necessary and certainly the correlation need not be linear.

This system of predictive verification was employed in the presentation of marmot sociobiology. Given the end points of long growing season-social intolerance (woodchucks) and short growing season-social tolerance (Olympic marmots), I predicted a correlation of intermediate growing season-intermediate social tolerance (yellow-bellied marmots). This was confirmed. In addition, predictive verification can be applied to other anticipated end-point correlations. Keeping with the marmot example, I predicted that if the described correlations were valid, then another marmot species also known to occupy one of the environmental end points should show comparable end-point behaviors. Hoary marmots were known to occupy high-elevation, short growing season environments, comparable to those of the Olympic marmots, but their social behavior was unknown. I predicted adaptations resembling those of the Olympic marmots, and this also was confirmed (Barash, 1974b; Plate 3.4).

Of course, confidence in the reality of a proposed correlation is not necessarily the same thing as proving the *cause* of the correlation, which we assume to be evolution by natural selection. The point of all this is to emphasize that any scientific hypothesis should be capable of being *disproved* as well as proved, and prediction therefore adds great power to the correlational approach. This leads us to the fourth proposed technique for employing evolution in studying behavior, the predictive approach. Of course prediction can often be applied to the historical and evaluative approaches as well, and in such cases it invariably leads to a stronger science. However, as described here, the predictive approach is at least somewhat distinguishable from the other three.

The predictive approach applied to social behavior is actually

PLATE 3.4 ADULT HOARY MARMOTS IN UPRIGHT PLAY-FIGHT. This behavior indicates a high degree of social tolerance and is best-developed among marmots inhabiting high-elevation, short growing season environments. (PHOTO BY J. TAULMAN)

novel and closely intertwined with the development of socio-biology itself. It relies on what may be called the Central Theorem of Sociobiology, a fundamental hypothesis that seems to underlie the discipline of sociobiology, although it has not been proposed explicitly until now. It states: When any behavior under study reflects some component of genotype, animals should behave so as to maximize their inclusive fitness.* In most cases this is achieved by maximizing the production of successful offspring. Most of the important, conceptually unifying notions of science tend to be both simple and of great potential power, and the Central Theorem of Sociobiology is no exception. It offers wide possibilities for a predictive approach to social behavior. With certain basic information concerning a species' biology, accordingly it should be possible to predict behavior patterns based on the assumption that these patterns will be such as to maximize the inclusive fitness of the performer.

For an example, consider the "male response to apparent female adultery in the mountain bluebird" (Barash, 1976a; Plate 3.5). Among monogamous species, the fitness of males would be greatly reduced if their mate copulated with a different male, since there is no evolutionary return in helping to rear another male's offspring. However, female copulations with other males should reduce male fitness only if it occurred at the time of possible conception; once the eggs are laid there are no evolutionary consequences of additional matings. Mountain bluebirds typically alternate domestic nest-side duties with brief foraging excursions, one mate remaining at the nest at all times. I experimentally affixed a stuffed museum specimen of a male near two mountain bluebird nests, so that when the mated male returned he discovered the apparently adulterous pair. I then recorded the behavior of the male after he made this discovery.

This experimental treatment was conducted at the time copulation normally occurs, 10 days later when the eggs were in the

---

*Inclusive fitness differs somewhat from Darwinian fitness, our concern in this and the preceding chapter, in that it considers the fate of genes rather than offspring, devaluing the importance of relatives in proportion as they are more distantly related. A treatment of inclusive fitness and its significance for sociobiology is found in Chapter 4.

PLATE 3.5 MALE (*right*) AND FEMALE (*left*) MOUNTAIN BLUEBIRDS OUTSIDE THEIR NEST BOX. (PHOTO BY H. POWER)

nest, and finally 10 days after that, shortly after the eggs had hatched. The Central Theorem predicts that the mated male would behave aggressively toward the model male, although with a progressive reduction in intensity and/or frequency with each successive discovery. It also predicts aggression toward the female; and further, that such aggression would be restricted to the presentations that occurred when female adultery could have resulted in offspring. These predictions were fulfilled (FIG. 3.5).

The high initial level of aggression toward the model is explicable in proximate terms by the presence of high levels of sex hormone during the breeding season. The progressive decline in such aggression is similarly explicable by regular seasonal declines in amounts of hormone. In ultimate terms, such a pattern is adaptive in that a male intruder at breeding time represents a maximum threat to the fitness of a resident male

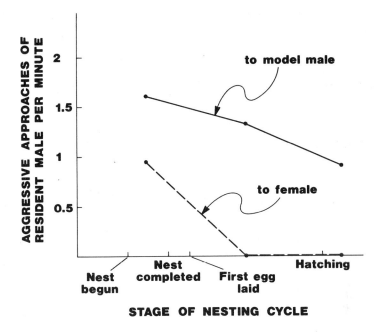

FIGURE 3.5. MALE RESPONSE TO APPARENT FEMALE ADULTERY IN MOUNTAIN BLUEBIRD. The resident male responded aggressively to the model male throughout the breeding cycle. In contrast, he responded aggressively to his mate only when the model was presented at the time copulation normally occurs. This pattern is in accord with predications that animals should behave so as to maximize their fitness. (*Modified slightly from Barash, 1976a.*)

because the chances of his inseminating the female are relatively high. As the season progresses, the danger posed by a stranger dwindles progressively, being reduced to that of a mere competitor for food resources. It is therefore in accord with prediction that aggression should also dwindle. In different proximate terms, the males' responses to the male models is explicable in terms of the ethological concept of releasers, in which animals are stimulated to perform instinctive behaviors by certain conspicuous, simple stimuli. Findings of this sort are familiar to ethologists and are not extraordinary in themselves. On the other hand, particular interest attaches to the male's response to his *female*.

The mated male behaved toward his female in accord with

the predictions based on fitness considerations; i.e., he was aggressive toward his mate when her potential copulations with a stranger would have threatened his own fitness. His tolerant response later in the year was also adaptive: Stolen copulations cannot hurt the resident male so long as the eggs have already been laid and his paternity is assured. Furthermore, because mountain bluebirds inhabit alpine environments and have a restricted breeding season, a male would almost certainly fail in re-nesting if he attempted to do so, even in the unlikely eventuality that he could induce a previously unmated female to attempt mid-season breeding. Such a male would be lowering his fitness by substituting a *possibility* of reproductive exclusion for a *certainty*. Published accounts of this research unblushingly employ the word adultery as it may well represent a true parallel to the human example (Lorenz, 1974). (See Chapter 10 for further attention to the controversial area of human sociobiology and the possible human implications of the above study and other fitness considerations).

Significantly, following the vigorous attacks of an enraged mountain bluebird male mate ("husband"), one female departed and was replaced by another female with whom the male successfully reared a brood. An aggressive male strategy of this sort would only be adaptive, i.e., maximize male fitness, if replacement females are readily available. It is not surprising that mountain bluebird populations at Mount Rainier appear to be limited by the restricted availability of nest sites, so there may well be a surplus of females from which the recently separated male can choose. Summarizing, male mountain bluebirds respond to apparent female adultery in a way that is consistent with maximizing their fitness. Males of this species that responded in this manner to similar situations in the past presumably produced more offspring than did those that behaved differently. An aggressive response to suspected adulteresses therefore became the predominant species-typical behavior, so long as jealous husbands could recoup their losses. In contrast, when replacements are less available, maximization of fitness would entail a different male strategy.

Many of the common North American puddle ducks (mallards, pintails, and shovelers, for example) are rapists. For all their innocent appearance, mallards are particularly notorious in this regard. Courtship behavior in mallards has been thoroughly documented by ethologists. It consists of a complex series of stereotyped behaviors and responses between the pair. Copulation, when it occurs, follows prolonged mutual interaction and obviously is agreed upon by both partners. By contrast, a male occasionally lands without warning on a mated female with whom he has *not* been courting; she resists, but the male nonetheless attempts to force a copulation. It seems justified to consider this as rape (Plate 3.6). And it is instructive to compare the response of the raped female's mate with that described for the seemingly cuckolded mountain bluebird.

As predicted, males attempt to intervene when they can, defending their reproductive rights by dislodging the would-be rapist. But when a male arrives too late, his response toward his

PLATE 3.6 TWO MALE MALLARDS RAPING A FEMALE (*closest to viewer*). Such behavior is common among this species, particularly when the female's mate is not present. When he is, his response is aggressive toward the would-be rapists, after which he often rapes the female himself. (PHOTO BY P. JOHNSGARD)

mate is also in accord with evolutionary prediction: He often proceeds immediately to rape the just-raped mate, a behavior not otherwise seen in mated pairs. In contrast to the situation among mountain bluebirds, there is generally a shortage of adult female mallards. Aggressive rejection of a raped female would therefore be unlikely to enhance male fitness, since he may well be unable to secure a new mate and would sacrifice his best chance of reproductive success. Under the circumstances, his best strategy should be to introduce his sperm as quickly as possible in order to compete with those of the rapist. He does this.

There are undoubtedly certain costs associated with being a web-footed rapist. First, there might be potential injury at the hands (wings?) of the defending male. Second, a rapist leaves his own mate susceptible to being raped while he is away. This suggests the further prediction that rapists among ducks should be more likely to come from the ranks of unmated males who have little to lose and much to gain, as opposed to mated males whose options are exactly reversed. This prediction is also confirmed.

Considerations of fitness provide the ultimate currency in which all behavior can be evaluated. This applies to both the research strategies of sociobiologists and to the way in which evolution operates on behavior. However, it is often difficult to measure fitness directly in terms of the production of offspring whose reproductive success should be evaluated. In most field studies, we must therefore often settle for indirect fitness measures such as the gross numbers of eggs laid, hatched, and/or fledged, offspring weaned, access to good-quality habitats, position in a dominance hierarchy, probability of avoiding predators, and efficiency in time budgeting and energy (food) harvesting. Considerations of this sort have been employed most elegantly in ecological studies of optimality modeling, in which behaviors are evaluated in terms of whether they represent optimum strategies, usually measured in terms of maximizing the difference between benefit and cost (Cody, 1974). To date, such studies have dealt most intensively with optimal foraging strategies (Schoener, 1971; Katz, 1974; Charnov,

1976), a topic with which we are not concerned directly, because it is somewhat removed from social behavior. However, fitness remains the essential referent in these treatments and is likely to reward similar approaches to other behavioral strategies as well.

The correlation of the above parameters with fitness is probably high, thus justifying the approach, but these behaviors are not *in themselves* of any evolutionary consequence, only fitness is. On the other hand, insofar as any behavior influences fitness and that behavior is influenced by genotype, then the behavior has consequences for its own evolution. And, the evolutionary process has consequences for the behavior.

When Alice was lost in Wonderland, she asked the Cheshire Cat:

"Would you tell me, please, which way I ought to go from here?"
"That depends a good deal on where you want to get to", said the Cat.
"I don't much care where—" said Alice.
"Then it doesn't matter which way you go", said the Cat.
"—so long as I get *somewhere*", Alice added as an explanation.
"Oh, you're sure to do that", said the Cat, "if you only walk long enough".

It may well be that like the Cheshire Cat, science cannot tell us *where* we should go. But once we have decided, it should be able to point out the best route. So far, I have attempted to show that a path leads from evolutionary biology to animal and possibly human social behavior. Whether we choose to follow is up to us.

F O U R

The Biology of Altruism

It is always nice to be nice to someone else. But it is unlikely to be adaptive. If helping another individual involves some cost to the helper, the helper is hurting itself, i.e., such behavior would be selected against relative to alternative, selfish behavior, unless there are compensating benefits that render helping ultimately advantageous to the performer. Moral injunctions are simply irrelevant to the evolutionary process. Insofar as animals possess genetically mediated tendencies to behave for the benefit of another, the logic of evolution demands that these tendencies be grounded in underlying selfishness. Otherwise, they would not persist.

In 1962, a Scottish ecologist, V. C. Wynne-Edwards, wrote a book titled *Animal Dispersion and Its Relation to Social Behavior*. Its central thesis was startling: Animals tend generally to avoid overexploitation of their habitats, especially with regard to food supply. They accomplish this largely by altruistic restraint on the part of individuals who reduce their reproduction, or refrain altogether, thereby avoiding local overpopulation. Furthermore, Wynne-Edwards proposed that much of social behavior itself functions as a mechanism whereby individuals are informed of their own numbers relative to the resources available, so that they can modify their reproduction accordingly. Considerable evidence was marshalled to support the notion of reproductive restraint. Thus, many animals seem capable of producing more offspring than they do. Often, subordinate members of a social dominance hierarchy do not reproduce. In addition, age at first breeding is frequently delayed beyond the point that appears necessary; seemingly inappropriate delays occasionally intervene between reproductive attempts; and parents occasionally even consume their own offspring.

Furthermore, it had long been recognized that animals engage in a strange variety of social behaviors, such as communal displays, winter roosting aggregations, or group vocalizing, for which no unifying explanation had been suggested. Wynne-Edwards proposed a single coherent interpretation for all these puzzling phenomena. At the same time he related them to the observed rarity of animal overpopulation, by seeing much of animal social behavior as *epideictic displays*, behaviors specifically adapted to provide an immediate local census, as a result of which individuals were then thought to employ the appropriate mechanisms of reproductive restraint. Even the daily vertical migration of oceanic plankton was interpreted in this manner, as were the pre-dawn choruses of howler monkeys in the Amazon rain forest and the large familiar flocks of starlings, blackbirds, etc. that enliven the evening skies of the United States. Although food was interpreted as the most common ultimate limiter of animal numbers, Wynne-Edwards suggested that animals employed conventional competition rather than direct competition for resources, thereby substituting these less costly social interactions in place of mortal issues. Accordingly, he proposed that animals made use of epideictic displays to inform themselves of local population levels and/or they employed such conventions as dominance or territorial maintenance, to avoid overpopulation with its attendant disadvantages.

Wynne-Edwards' thesis is appealing, not only because it provides a coherent world-view (a paradigm of sorts) but also because it corresponds to our recognition of the potential conflict between individual and societal benefit. Thus, in a compelling essay on a wide-ranging human problem, Hardin (1961) discussed the "tragedy of the commons". The English commons was land in public ownership, where sheep grazed. The tragedy was that each participating shepherd perceived that he derived maximum benefit by having his sheep graze as much as possible; otherwise someone else would use up the available forage. Of course, the inevitable consequence was a drastic deterioration of the commons—due to over-grazing. Altruistic restraint was not profitable to individuals, so each followed a selfish policy as a result of which all suffered. Similarly, people now recognize that commonly "owned" whale populations have been decimated by the selfish pursuits of individual nations. Some form of societal regulation, as ostensibly provided by the International Whaling Commission, is essential for preservation of whale fisheries. But it is another question whether some equivalent to this system can have evolved within free-living animal societies, in which selection presumably maximizes individual fitness, not group benefit.

The obvious desirability of such controls from a human viewpoint may explain the attractiveness of Wynne-Edwards' proposals. But the ideal should be distinguished from the real. Although Wynne-Edwards was not the first to propose the occurrence of natural, altruistic restraints upon reproduction, his book summarized an enormous amount of data and was a monumental achievement. However, it also brought about thoughtful criticisms, based on both theoretical and empirical issues. On the empirical side, it was pointed out that each social behavior pattern identified as epideictic has been explained more parsimoniously in other, more specific ways: Howler monkeys howl to mark and defend their territory (Carpenter, 1934), communal courtship displays increase the likelihood of attracting females (Snow, 1963), etc. Furthermore, available evidence suggests that reproductive performance is adjusted to result in the maximum production of surviving young, with no evidence of altruistic restraint (Lack, 1968). In other words, what may appear to be a submaximum reproductive performance may well represent the maximum for each animal.

For example, the English swift (a small bird) typically lays two eggs per clutch. Some females, on the other hand, produce three or four eggs, suggesting that two-egg individuals are engaging in a submaximal strategy, perhaps to prevent overpopulation. However, egg production in itself is only important insofar as it results in the production of successful offspring. Further research revealed the following relationship (data from Perrins, 1964) between number of eggs laid and number of young successfully fledged (reared to independence).

Clutch size	Fledglings (%)	Fledglings per nest (mean)
2	82	1.64
3	45	1.35

Apparently, the presence of too many offspring reduces the food available to each, with the result that parents of excessively large clutches actually produce fewer offspring than parents who lay fewer eggs but provide more nourishment for each. The deviation in both directions from the optimum clutch size of two is probably due to unavoidable mutations and/or recombinations, which provide genetic variability from which selection could generate a new modal clutch size if food conditions either improve or deteriorate (Plate 4.1).

Despite the cogency of such empirical findings, the theoretical questions were of greater importance both for the issue at hand and for the future development of the theory of altruism and sociobiology itself. Wynne-Edwards recognized that his scheme posed a problem for evolution by natural selection, a problem that generalizes to other altruistic behaviors, beyond simple reproductive restraint. Imagine a gene R that induces its possessers to express reproductive restraint, while gene r causes a selfish insistence on maximum reproduction. How can R persist and spread in a population when its effect is to reduce its representation in succeeding generations? In other words, gene R carries an automatic selection against itself relative to gene r, because carriers of r will by definition leave more offspring than carriers of R. Insensitivity to epideictic behaviors should therefore be selected and/or a refusal to abide by the various altruistic social conventions postulated above. The basic problem then is how to explain the evolution of a self-defeating tendency. In our example, what could stop altruistic R from being swamped by selfish r?

PLATE 4.1 ADULT ADELIE PENGUIN FEEDING A CHICK WHILE A SECOND ONE STANDS BY. This species typically lays two eggs; in years when food is abundant, both survive; when food is scarce, only the older one survives. (PHOTO BY R. TENAZA)

Group Selection

Wynne-Edwards' proposed mechanism has been labeled *group selection*; understanding it and its limitations is essential to a grasp of sociobiology. If we grant that individuals normally exist within distinct social groups, then reproductive restraint (R) can evolve if its presence confers sufficient reproductive advantage upon the group as a whole. Similarly, reproductive selfishness (r) must confer a strong disadvantage to the group in order for altruism to evolve through group selection. Thus, natural selection acting at the level of groups has been invoked as a mechanism for the evolution of traits that are individually disadvantageous but beneficial to a larger social unit.

Group selection was controversial from the start, with most biologists arguing forcefully against it (e.g., Crook, 1965; Wiens, 1966; Williams, 1966a). Natural selection was already known to operate on individuals and any phenomena attribut-

74

able to group selection were explained more parsimoniously in other ways. Furthermore, selection operating on individuals *within* groups would almost certainly act more strongly to eliminate altruistic traits than selction *between* groups could act to maintain these traits. On the other hand, there was some support for group selection. For example, certain tropical birds were found to be capable of rearing more offspring than they normally do (Skutch, 1949), i.e., when additional young were placed in the nest, parents were often able to rear all of them.

Recently, computer models have been developed in an effort to describe the exact conditions under which group selection can occur, at least in theory (Levins, 1970; Boorman and Levitt, 1972, 1973; Gadgil, 1975). Specific results obtained varied with the initial assumptions, but a few generalizations are emerging. Thus, in order for group selection to occur favoring individually disadvantageous traits, the extinction rate of *groups* would have to be comparable to that of *individuals* within the groups; i.e., differential reproduction of groups must equal or exceed differential reproduction of individuals within groups. This is because altruists would be losing out to more selfish individuals within each group, so that their persistence in the population as a whole would accordingly require groups containing altruists to reproduce much more successfully than those lacking altruists. The presence of altruist genes would have to contribute dramatically to the differential reproductive success of groups, such that the relationship between gene frequency and group extinction rate approaches a step function (FIG. 4.1). Finally, group selection requires that there be very little genetic exchange between groups (emigration and immigration) so as to miminimize the likelihood of altruistic genes being exposed to within-group competition from newly arrived selfish genes.

Even with these requirements satisfied, altruistic genes would not remain stable within any one group; given enough time, they would be replaced (selected against) by competing selfish genes within each group and would therefore exist among the species as a whole only if the groups of which they are members reproduce fast enough so that the altruist genes are replicated in new groups before they disappear in the parent groups.

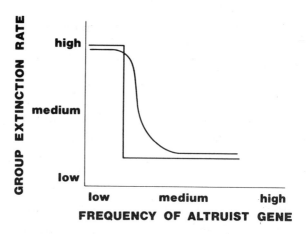

FIGURE 4.1. THE NECESSARY RELATIONSHIP BE-
TWEEN FREQUENCY OF ALTRUIST GENE AND GROUP
EXTINCTION RATE IN ORDER FOR ALTRUISM TO
EVOLVE BY GROUP SELECTION. The assumption is
based on Boorman and Levitt (1973). The relationship may be a step function or a logistic curve, but in
either case increase in the frequency of the altruist
gene must cause a dramatic decrease in group extinction rates if altruism is to be selected at the level of
groups.

Thus, the present concensus among theoretical ecologists and
evolutionary biologists seems to be that the evolution of altruism by group selection is theoretically possible but that its occurrence is very unlikely given the low probability of satisfying
all of the conditions that appear to be necessary (see Wilson,
1973, for a readable account of recent mathematical treatments; Williams, 1971, has edited a though-provoking collec-
tion of more advanced essays; and D. S. Wilson, 1975, has provided a novel mathematical formulation of possible requirements for group selection).

There is one case in which group selection is generally acknowledged to have occurred, and consideration of this example in some detail may further clarify our understanding of the
apparent requirements for group selection (Lewontin, 1970).
When rabbits were introduced to Australia, they quickly became major agricultural pests. Their numbers were finally controlled by introducing the myxoma virus, which causes the dis-

ease myxomatosis, into the wild population. Interestingly enough, an evolutionary change occurred such that the virus actually became less virulent over time. This is measured by the lethality caused by a virus; the lethality is itself a function of reproductive rate. Thus, the evolution of reduced virulence of myxoma viruses involved an altruistic decrease in reproductive rate among the viruses, as follows: The likelihood of one rabbit infecting another, via a mosquito who bites them both, varies directly with the lifespan of the infected animal. Lifespan, in turn, varies inversely with the reproductive rate of the myxoma viruses that inhabit each rabbit. Rabbits infected by virulent viruses have a shorter lifespan than rabbits that harbor less virulent forms. Accordingly, rabbits suffering from a virulent infection were less likely to infect other rabbits than were those whose viruses reproduced less often. Therefore reduced reproduction evolved among the viruses.

In this example, the rabbits constituted group boundaries for each inhabiting population of viruses. Nonvirulent viruses were altruistic, relative to virulent ones, because they exhibited reproductive restraint. This restraint decreased the immediate reproductive success of its carriers and was presumably selected against *within* any one rabbit. On the other hand, reduced virulence contributed to group survival and reproduction (the spread of the disease from one rabbit to another) and was, therefore, positively selected at the level of groups. The requirements of group selection are starkly apparent in this example: Groups (viruses contained within a single rabbit) are discrete entities, between which extinction rates are high enough to counter the lower reproductive success of individual altruists within any one group. Under these conditions, group benefit can outweigh individual benefit.

If the reader balks at the notion of an altruistic virus, it must be emphasized that the term altruism is defined here solely by its consequences: an act that reduces the personal reproductive success of the performer while increasing the reproductive success of others. As with other behaviors encountered in sociobiology, no implication is made regarding conscious awareness on the part of the performing individual, in this case the altruist. All that is required is that individuals of one genotype with a tendency to behave in a particular way, perhaps in conjunction with particular environmental circumstances, leave more offspring than do other individuals whose genotype induces them to behave in other ways. In this sense, there is nothing incongruous about altruistic viruses, turkeys, or honeybees, so long as the behavior fits the definition and an appropriate ultimate mechanism can be demonstrated.

Up to this point, our consideration of altruism has been limited to overt reproductive restraint, but it need not be. Many behavioral tendencies, especially those reflected in social behavior, are relevant to the biology of altruism. Thus, an individual may share food, territory, or nest with another. It may help protect another from predators, either by giving an alarm call or by overt defense. It may help care for another's offspring and/or refrain from damaging another animal during aggressive encounters. Through a wide range of solicitous behavior, an animal may assist the survival and/or reproduction of another, often at some cost to itself; if nothing more, the time spent in aiding another could be put to more selfish use. In most cases, however, the cost is more obvious: increased likelihood of being taken by a predator, less food consumed because some is shared, etc. Finally, any behavior increasing the success of a competitor has consequences, usually negative, for the altruist.

Ultimately, of course, all these behaviors are related in that they influence the personal fitness of the animal concerned, i.e., they all reduce to various patterns of reproductive restraint. For example, imagine a behavior whose effect is to decrease the net calories available to an animal, perhaps because that animal vocalizes when it discovers food, thereby informing others of the location of the food. The evolutionary calculus operating on this behavior evaluates it insofar as fewer calories translates into fewer offspring. This may occur, either directly or through the greater likelihood of physical exhaustion and the consequent predation or disease, brought about by the greater risk associated with increased feeding time, distance, or effort. Just as money is the root of all evil, fitness is at the root of all evolutionary considerations and altruism of any sort is an evolutionary conundrum when it appears to reduce fitness.

True altrusim, in the sense of giving more than one gets, should therefore never evolve, because individuals demonstrating such behavior would be, by definition, less fit than their selfish competitors. And yet, altruistic behaviors are surprisingly common among animals, as well as among humans. Group selection appears to be inadequate as an ultimate explanation. The remainder of this chapter will consider various alternate routes for the evolution of altruism, based essentially on models of selection operating at the level of *individuals*.

Kin Selection

One of the most cogent explanations of the biology of altruism goes by the name of *kin selection* (Maynard Smith, 1964). This concept actually derives from the pioneering theoretical work of the British population geneticist W. D. Hamilton (1964), who originally applied his model to explain the evolution of complex social systems and reproductive restraint in insects. Basically, Hamilton recognized that the production and care of offspring is only a special case of solicitude for other individuals with whom one shares genes through common descent. Thus, one might legitimately ask: Why reproduce? Breeding represents a large expenditure of both time and energy. Parents must nourish their young; provide a safe rearing place; suffer increased risk to predators during courtship; provide and defend their offspring; and generally lavish time and attention which is therefore unavailable for themselves (Plate 4.2). It is costly to reproduce, and there is considerable evidence that the burden of breeding takes its toll in reducing parental lifespan (Taber and Dassman, 1956).

Yet, all organisms reproduce. Why? Like most important questions, this one is both trivial and profound. All living things are themselves the products of parents who reproduced, and it is clear that of all phenotypic traits, those directly relevant to

PLATE 4.2 ADULT FEMALE DEER MOUSE RETRIEVING PUP TO HER NEST WHILE OTHERS IN HER LITTER ATTEMPT TO SUCKLE. (PHOTO BY R. ROBBINS)

the production of offspring would have the strongest influence on fitness. A genetically mediated tendency for childlessness would clearly not have a bright evolutionary future! The biological imperative of reproduction is universally strong because in ultimate terms offspring represent the primary means of leaving genetic representation in the next generation. For sexually reproducing, diploid species (those possessing two sets of chromosomes, one from the mother and one from the father), parents share one-half of their genes with each of their offspring. Sperm carry one-half of the father's genes and eggs carry one-half of the mother's; upon fertilization, these merge to form a new individual whose wholeness represents one-half of the genotype of each parent. This is why parents are concerned with the production of offspring. But at the same time, it suggests the importance of relatives and the sense in which offspring are only a special case of relatives in general: Natural selection operates by differential reproduction of *individuals*, but ultimately this is translated into competition among *genes*

80

and the closer the relative, the higher the proportion of shared genes.

Some examples will help explain. Consider the case of alarm calling in which, for example, a prairie dog utters a loud call or alarm when it sees a coyote approach the "town." In doing so, the alarm giver probably draws attention to itself, increasing the likelihood that it will be captured by the predator. Of course, by warning fellow prairie dogs who might not otherwise have known of the coyote's approach, the alarmist has improved the probability that these others will survive. Hence, it has increased *their* personal fitness while suffering apparent decrement to its own. In this sense, it is behaving altruistically. An alternative, selfish strategy would have been to retreat quietly and safely into a burrow and allow its unsuspecting colleagues to suffer the consequences. Why isn't this done? Insofar as the alarmist benefits its own offspring, such behavior poses no special problem. It is simply another example of parental martyrdom. The crux of kin selection theory is that the same argument applies for any beneficiary of the behavior, so long as the beneficiary is in some way related to the altruist, with the beneficiary devalued in proportion as that relationship is genetically more distant. Kin selection theory is an important modification of the more traditional, natural selection. According to kin selection, animals maximize their *inclusive fitness*—their net genetic representation in succeeding generations, including other relatives in addition to offspring.

The question then: Assuming that alarm calling reduces personal fitness, how can it evolve? Postulate a gene for alarm calling, C, and an alternative gene for selfish silence, c. If possessors of gene C are more likely to be eaten by sharp-eared coyotes, then possessors of gene c will clearly be more fit, and gene C (i.e., alarm calling) will disappear. Alternatively, gene C could theoretically be maintained by group selection if groups with alarmists were much more successful than groups without. But the group benefit would have to be very great, and in any case gene C would be replaced rapidly by gene c *within* each group. These difficulties of group selection were considered above.

Now, let us consider the kin selection argument: Gene C could persist in the population if its ultimate effect is to maintain or increase the net representation of other copies of gene C in the next generation. Thus, if the alarm calls produced by an individual carrying gene C increased the fitness of sufficient numbers of other individuals also carrying gene C, then gene C could be selected, *even though its effect on the bearer is to reduce its own personal fitness.* Insofar as the evolution of alarm calling or any other altruistic trait is concerned, the important point is the fate of *genes* coding for the behavior in question, not the fate of the *individuals* performing that behavior. Alarm calling could therefore be selected if the caller alerted a sufficient number of relatives. In this case, the required number of beneficiaries would decrease as their relatedness to the altruist increases; as

PLATE 4.3 OLYMPIC MARMOT ALARM CALLING. These animals live in closely-organized social groups within which genetic relatedness is high, i.e., there is a high probability that colony members share genes by virtue of common ancestry. Therefore, a genetically-mediated predisposition for alarm-calling could be maintained even if it results in increased mortality for the alarm caller. (PHOTO BY J. SPURR)

the proportion of genes in common increases. In other words, alarm calling would be selected if it resulted in the saving of a small number of close relatives or a large number of distant relatives. Significantly, there is good reason to believe that most animals tend to live in social groups within which relatives are more likely to be close together, regardless of whether the exact degree of relatedness is recognized (Plate 4.3).

Hamilton's original formulation dealt especially with the *eusocial* insects, those species possessing a specialized, nonreproductive worker caste, such as honey bees, ants, and wasps. These animals pose a unique problem for evolutionary theory, since their workers have foregone reproduction themselves, while they labor for the reproductive success of their mother, the queen. Eusociality is therefore a true pinnacle of altruism. Significantly, all eusocial species (except the termites) are haplo-diploid, i.e., the males are haploid (have one set of genes), and the females are diploid (have two sets). Hamilton pointed out that this genetic peculiarity results in female workers sharing three-quarters of their genes with their sisters as opposed to only one-half of their genes with their own offspring, if the workers were to reproduce (FIG. 4.2). As a consequence, an individual worker does more to further her evolutionary fitness by staying home to care for her sisters than if she were to leave the hive and attempt to raise a family of her own! Furthermore, the notorious laziness of the drones is entirely in keeping with their low genetic interest in the rest of the hive; they share only one-fourth of their genes with their sisters.

The social insects have had a major role in the development of sociobiology. For a thorough treatment of these fascinating animals, see Wilson (1971). The application of Hamilton's kinship theory has been criticized by other biologists, who emphasize the possibility that queens may be inseminated by several males, thus reducing the genetic relatedness of sisters (Alexander, 1974). The point is that if workers are only *half-sisters* (having different fathers), then their relatedness would be reduced from three-fourths to one-half, and inclusive fitness would not be maximized by helping to rear more workers, rather than reproducing personally. Recently however,

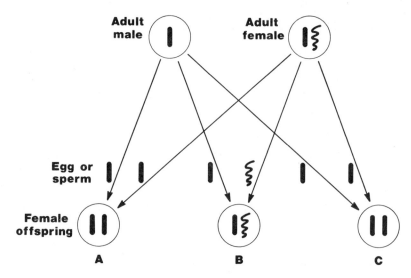

FIGURE 4.2. The Genetic Consequences of Haplo-Diploidy in the Eusocial Insects. The situation for one pair of chromosomes in bees is shown. Females are diploid so the adult female, the queen, has two such chromosomes. Males are haploid so the adult male has only one. He therefore contributes the same genetic material to each of his offspring and because this constitutes one-half of their genotype, each of his offspring must share *at least* one-half of their genes. Females may contribute either chromosome to each of their offspring. If two offspring receive the same chromosome from their mother, they are genetically identical with regard to that chromosome pair (Offspring A and C). If they receive different chromosomes, then they still share one-half of their genes (Offspring A and B). Because many chromosomes are involved, female offspring share on the average three-fourths of their genes; by contrast, if one of these females attempted to breed, she would share only one-half of her genes with any offspring produced, just as in this diagram the adult female shares only one-half of her genes with each of the offspring shown.

important support has been provided for Hamilton's theory by the demonstration that workers provide about three times the food for their sisters (other workers) as they do for their brothers (the drones), consistent with their three-fourths versus one-fourth genetic relationship (Trivers and Hare, 1976). For other important treatments of insect sociality, see Lin and Michener (1972), Michener (1974), Michener and Brothers (1974), Eberhard (1972), and West-Eberhard (1975).

Regardless of the cogency of kin selection applied to the social insects, the concept is of enormous explanatory power and is basic to the intellectual momentum of sociobiology. In fact, the field can be considered to have originated with Wynne-Edwards' treatment of animal social behavior and group selection (1962) and the reevaluation of natural selection it inspired among biologists (see Alexander, 1975), combined with Hamilton's development of kin selection theory. The interested reader should consult Hamilton's occasionally difficult, but always stimulating, papers on the subject (1964, 1971, 1972, 1975). Incidentally, for a third milestone in the currently unfolding history of sociobiology, we might identify the superb synthesis by E. O. Wilson, published in 1975. Two other watersheds are the theoretical contributions of R. L. Trivers (1971, 1972, 1974; to be discussed later), and the demonstration of adaptive function in animal social organization, as described by Crook (1970) and pursued by numerous investigators.

Kinship theory is probably applicable to all species and is certainly not limited to the social insects. Perhaps its central formulation is that genes for altruistic behavior will be selected if $k > 1/r$, where k is the ratio of recipient benefit to altruist's cost, and r is the coefficient of relationship between altruist and recipient, summed for all recipients. Further understanding of the theory requires separate consideration of each of these factors. The coefficient of relationship, r, is the proportion of genes in two individuals that are identical because of common descent. Thus, r between identical twins (all genes in common) is 1, while r between totally unrelated individuals is 0 (although in truth it is probably always above 0, since no two individuals in the same species can be absolutely unrelated.) When the genetic relationship between two individuals is not known, the above equation should be amended to read $k > 1/\bar{r}$, where $\bar{r}$ is the *average* r between the altruist and all other members of the same species with which the altruist has any possibility of interacting.

The r between parent and offspring is ½; between full siblings, ½; half sibs, ¼; uncles (or aunts) and nephews (or nieces), ¼; cousins, ⅛; etc. In the general case for diploid species without inbreeding, r equals $(½)^L$, where L is the number of generation links between the two individuals concerned. When more than one genetic path exists between the two individuals, $(½)^L$ is

85

summed for each path (FIG. 4.3). (See Li (1955) or Crow and Kimura (1970) for a more detailed treatment.)

The higher the r between two individuals, the more genes in common. Back to $k > 1/r$: for large r's (close relatives), $1/r$ is relatively small, so that k need not be large in order for altruistic behavior to be selected. In other words, the benefit to cost ratio can be small—the benefit derived by the recipient need not be very great in order for altruism to be selected, so long as the recipient is a close relative. Similarly, when r is large (the recipient is a close relative), altruism will be selected even if the cost to the altruist is comparatively great; i.e., even with a large denominator, k (benefit/cost) can still be greater than $1/r$ so long as r is large (because $1/r$ is then small). So, individuals should willingly run a greater risk to help a close relative (high r) than a more distant one (low r). On the other hand, when r is small (the recipient is only distantly related) then $1/r$ is large and in order for altruism to be selected the recipient's benefits must be great and/or the altruist's cost must be small.

Next, let us also briefly consider the factors which influence k (the ratio of benefit to cost). Benefit to recipient is increased by recipient need, e.g., assistance to an individual in mortal danger should be more readily given (because it will have been selected), all other things being equal. Aside from need, some individuals may particularly be able to profit from assistance rendered. Such "super-beneficiaries" (West-Eberhard, 1975) may include eusocial insect queens, which are specialized egg-laying machines, or adult birds, which have succeeded in defending hotly-contested territories and/or constructing elaborate and difficult nests, etc. In general, relatives should be predisposed for increased altruism toward recipients who possess high reproductive potential, as indicated perhaps by physical condition, among other things. Interestingly, among *Polistes* wasps in which several females may be simultaneously reproductive within a single hive, deterioration of the environment (forcing some of the reproductives to become altruistic, non-reproductive workers) induces those females with the smallest ovaries to behave altruistically. Animal antagonism toward aberrant individuals may be particularly significant in this re-

Full siblings　　　　　　**Half siblings**

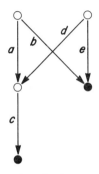

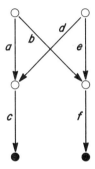

Aunt-niece　　　　　　**Cousins**
(or uncle-nephew, etc.)

FIGURE 4.3. COEFFICIENTS OF RELATIONSHIP (r) FOR
VARIOUS RELATIVES. Each arrow introduces a 50%
probability that the two individuals thus connected
share genes. Therefore, the likelihood that any particular gene "gets through" several such arrows is
simply 0.50^L, where L is the number of arrows. When
two individuals have more than one ancestor in common, e.g., full siblings, then they can share genes via
both; therefore we must add all the possible 0.50^L's.
For full siblings, $r = (a \times b) + (c \times d) = (0.50 \times 0.50)$
$+ (0.50 \times 0.50) = (0.50)^2 + (0.50)^2 = 0.25 + 0.25 =$
0.50. For half siblings (different fathers), $r = (a \times b) =$
$(0.50 \times 0.50) = 0.25$. For aunt-niece pairs, $r = (a \times b$
$\times c) + (d \times e \times c) = (0.50 \times 0.50 \times 0.50) + (0.50 \times$
$0.50 \times 0.50) = (0.50)^3 + (0.50)^3 = 0.125 + 0.125 =$
0.25. For cousins, $r = (a \times b \times c \times f) + (d \times e \times c \times f)$
$= (0.50 \times 0.50 \times 0.50 \times 0.50) + (0.50 \times 0.50 \times 0.50$
$\times 0.50) = (0.50)^4 + (0.50)^4 = 0.0625 + 0.0625 =$
0.125.

87

gard, as well as the nearly-universal human tradition of infanticide directed toward deformed new-borns. (See Chapter 10 for specific treatment of the possible relevance of kin selection and the biology of altruism to human behavior.)

The denominator of k (cost to altruist) may also be influenced by several factors. Low risk behaviors should be more likely to evolve than high risk behaviors. The cost of altruism may be low if the individual is highly adapted to provide assistance, e.g., if it is powerful and well-armed, and therefore less likely to suffer because of the assistance it provides (perhaps defense against a predator or competing group). Similarly, such an individual is probably also likely to provide more benefit to those aided. In addition, just as altruism is increasingly likely in proportion as the beneficiary can in fact benefit, it is also more likely in proportion as the altruist is losing little; i.e., altruism should vary inversely with the altruist's personal reproductive potential. If that potential is low, for example, because of infirmity or senescence, then the cost to the altruist is correspondingly low and altruism should be more likely. From evolution's viewpoint, in proportion as an individual has little to gain (reproductively) by living, it has little to lose by dying.

The upshot of kinship theory is an expanded view of fitness, recognizing gene frequency as reflected in all relatives, rather than simply the production of offspring. *Inclusive fitness* is the term used to incorporate the summed consequences of both personal fitness (via offspring) and fitness derived via the representation of genes in relatives. Behavior, whether altruistic or not, will evolve if and only if it is mediated by genes whose effect is to increase the inclusive fitness of the bearer relative to the consequences for inclusive fitness of alternative behavior, mediated by alternative genes. It provides a coherent theory for the biology of nepotism among living things (Plate 4.4).

Kinship theory and inclusive fitness apply directly to the real world. For an extreme and simplified case, an individual would be selected to sacrifice its life to save its siblings only if it could save more than two of them (making $k > 1/(\frac{1}{2})$). The implications of kinship theory for parent-offspring relations are explained in Chapter 7. Several basic predictions concerning altruistic behavior derive from kinship theory and all have been confirmed in field observations (Brown, 1975): (1) Altruism is more common among closely related than among distantly related individuals; (2) Altruism is more common among species with relatively little dispersal that therefore are more likely to share kinship ties with their neighbors; and (3) Species exhibiting altruism are relatively more likely to discriminate against outsiders (nonrelatives) and to recognize insiders as individuals, thereby increasing the likelihood that they can behave differentially as a function of differential r's.

Adult zebras defend calves against predators, whereas wildebeests do not. Significantly, zebras live in coherent family groups whereas wildebeest herds experience substantial mixing of genetic lineages, making it unlikely that an adult is related to a randomly chosen calf (West-Eberhard, 1975). Among North

PLATE 4.4 THE TWO ADULT MALE LIONS IN THE BACKGROUND COOPERATED WITH EACH OTHER TO GAIN CONTROL OF A PRIDE FROM THE DOMINANT MALE IN THE FOREGROUND. Significantly, the two cooperating animals were brothers. (PHOTO BY K. EATON)

American wild turkeys, brothers vie for social dominance within brotherhoods composed of two males each. When established, these brotherhoods compete with other, similar pairs for dominance within a group. Then, groups compete for dominance within the local population, after which the dominant male within the dominant duo within the dominant group does most of the mating. Within each brotherhood subordinates assist in competition with other duos, and, in addition to defering to the dominant, they even help attract females with which they do not mate (Watts and Stokes, 1971; Plate 4.5). In terms of inclusive fitness, each male would be better off if it fathered offspring, and, accordingly, males compete vigorously for dominance within each brotherhood. Once a male has lost, however, his next-best strategy is to attempt, altruistically, to further the reproductive success of his brother. Although he has lost the chance of becoming a father, he may at least be an uncle and gain inclusive fitness over the alternative strategies of going it alone, sulking in the corner, or whatever. Simultaneous polyandry or wife sharing is very rare among animals. Significantly, in the only other case in which it has been described, the Tasmanian native hen, the cooperating males are also brothers (Maynard Smith and Rid-

PLATE 4.5 A BROTHERHOOD OF WILD NORTH AMERICAN TURKEYS COURTING A GROUP OF FIVE FEMALES. Within such brotherhoods, competition for dominance is intense; however, after the issue is decided, the subordinate contributes to the reproductive success of his brother, even though he will not breed himself. (PHOTO BY D. SMITH)

path, 1972; Plate 4.6). It may also be of interest that among human societies that practice polyandry, fraternal polyandry is quite common (see Chapter 10).

Kinship theory may be tested with the use of *quantitative* ecological data, supplementing the *qualitative* observations described above (Brown, 1975). Thus, it is common among certain species of birds for young, sexually competent adults to assist other, older adults in rearing their offspring, rather than attempting to breed themselves (Skutch, 1935, 1961; Fry, 1972). Significantly, in most cases these helpers at the nest are aiding their parents and hence gaining inclusive fitness by rearing siblings. We can consider that a potential helper has two available options: Stay with its parents and help rear siblings, the *helper* strategy, or leave the nest and attempt to breed, the *selfish* strategy. For selection to favor helping, that strategy must contribute more inclusive fitness than the alternative, self-

PLATE 4.6 THREE ADULT TRUMPETER SWANS CARING FOR A SINGLE BROOD OF YOUNG. Associations of this sort are common among swans and geese. Their significance is unclear. However, it would be interesting to know whether the third adult is a sibling of one parent, thereby gaining inclusive fitness by helping to rear nephews and nieces. Alternatively, it could also be an offspring from a previous year, gaining inclusive fitness by helping to rear siblings. (PHOTO BY D. PAULSON)

ish strategy; $H \times r_H$ must be $> S \times r_S$, where H is the number of helped siblings brought to maturity *beyond* what the parents would produce without aid; i.e., H is the additional, inclusive benefit accruing to helpers by virtue of their helping; r_H is the coefficient of relationship between the helpers and the offspring they help rear; and S is the number of offspring raised by selfish individuals, assuming no helpers; and r_S is the coefficient of relationship between the selfish parents and their offspring.

If the helpers are helping to rear full siblings, $r_H = \frac{1}{2}$, as does r_S (r between parents and offspring), so the requirement for altruistic helping simplifies to $H > S$. Among the Florida scrub jays, parents without helpers rear 0.5 independent young per nest (S), whereas parents with helpers rear 1.3 (Wolfenden, 1975). Therefore, $H = 1.3 - 0.5 = 0.8$ and helping should evolve, because $0.8 > 0.5$, i.e., $H > S$. As expected, helping commonly occurs in this species.

Among a tropical species, the "superb blue wren," parents without helpers rear 1.5 independent young (S) and parents with helpers rear 2.83 (Rowley, 1965). Therefore, $H = 2.83 - 1.5 = 1.33$, and helping should not evolve, since $1.33 < 1.5$, i.e., $H < S$). In fact, females of this species do not help. There is a relative shortage of female wrens and therefore only about 70% of the males are able to breed. Adjusting the value of S for males accordingly, we get $S = 1.5 \times 0.70 = 1.05$, and of course $1.33 > 1.05$, so for males, $H > S$. Therefore, males in this species should help, and they do. However, each male either breeds or he does not; he cannot receive a 70% return, so his options during his first selfish breeding attempt are really 1.5 or 0. Females are maximally fit if they breed; males are also maximally fit in the same manner but only if they are successful in obtaining a mate. Therefore the optimal strategy for males is to seek to acquire a mate, but to become a helper if unsuccessful; and this is what they appear to do.

It is also interesting to consider the factors influencing the strategic decision of when to become a selfish parent rather than a helping sibling among Florida scrub jays, for example, in which the former decision provides a relative payoff of 0.5 and the latter, 0.8. Since parenting is less profitable than helping, why do any individuals attempt the former? Remember that a parent *with helpers* enjoys a return of 1.3 offspring. So, even though the transition from helper to parent may involve a short-term decrement in inclusive fitness, exchanging 0.8 for 0.5, in the long run it is profitable or maximizes inclusive fitness, because in subsequent breeding seasons these repeat parents will have their own offspring of the previous year as helpers, thus garnering 1.3 successful offspring from then on. Some readers may be uncomfortable with the notion of animals doing their arithmetic and then behaving accordingly. Recall that such computations are not required of the animals themselves. Rather, animals have been selected for behaving in certain ways under certain circumstances. The arithmetic was done by natural selection: Individuals who behaved in a way that produced maximum return (measured in units of fitness) were positively selected. Therefore the species came to be composed of individuals each of whom behaved *as though* it knew its arithmetic (Plate 4.7).

Kinship theory implies that the interest of one animal in the well-being of another should vary directly with the proportion of genes shared by virtue of common descent. Assistance should therefore be unlikely in the absence of genetic relatedness. This hypothesis was tested by removing one mate each from twenty-five parental pairs of nesting mountain bluebirds. In ten cases, the eliminated individuals were replaced by new consorts, who probably were not closely related to the nestlings that had already been produced; in any event they were almost certainly less closely related than the original parents. Significantly, only one of these "foster parents" fed the nestlings, and none gave alarm calls under conditions of danger (Power, 1975). Presumably, these replacement adults were consorting with the new single parents because of the potential of producing a *subsequent* brood, in which they would have the usual, parental stake. But until then, investing in unrelated offspring is a low fitness strategy and therefore did not occur.

PLATE 4.7 AN ADULT FEMALE BLACK-BACKED JACKAL (*center*) IS ABOUT TO REGURGITATE FOOD TO AN EAGER GROUP OF LARGE JUVENILES, NOT HER OFFSPRING. Although rare in mammals, helping by non-parents does occasionally occur. Among black-backed jackals, non-parents are responsible for 30% of the food regurgitated to young; unfortunately, genetic relatedness between helpers and beneficiaries is not known in this case. (PHOTO BY P. MOEHLMAN)

Reciprocity

Despite its enormous potential and utility, kinship theory is not unique in explaining altruistic behavior within the traditional framework of natural selection acting to maximize the fitness of individuals rather than groups. A major alternative mechanism is *reciprocity,* an appealing notion that can select for altruism with no assumptions whatever concerning genetic relatedness (Trivers, 1971). Indeed, it could even operate between members of different species. The basic requirement in order for altruism to evolve via reciprocity is that the performance of altruistic behavior must result in a return of altruistic behavior toward the original altruist such that the ultimate benefit in units of inclusive fitness is greater than the cost. In other words, the return to the altruist must be greater than the decrement in inclusive fitness imposed by the original altruistic act. It is a sort of biological Golden Rule, because the reciprocation would presumably come from the original beneficiary.

As a practical matter, certain conditions would have to be met in order for reciprocal altruism to be a viable system. The altruistic act should carry a low risk for the altruist and confer a high benefit for the recipient, because the *origin* of reciprocity would rely largely upon chance and would therefore be unlikely if the system was too heavily weighted against the first few altruists. Furthermore, there would have to be a high probability that the situation would be reversed such that the altruist would be in a position to benefit by low-risk altruism from the original recipient. And the original recipient would have to be able to recognize the altruist as an individual, so as to reciprocate appropriately; otherwise, altruism might go unreciprocated, which in itself would constitute a strong selective disadvantage to the original behavior. In other words, unless the altruist gets enough back in return for its cost, such behavior will not evolve.

These considerations suggest further that reciprocity is most likely to evolve among intelligent, closely integrated social species in which the opportunities for reciprocity and individual recognition would be greatest. Finally, they suggest one of the major factors working against the evolution of reciprocal altruism: the danger of cheating. Reciprocal systems are highly vulnerable to the evolution of cheaters, individuals who accept the altruism of others but fail to reciprocate when the occasion arises. This strategy would be viable as it enabled individuals to enhance their personal fitness via altruism received without losing any by the subsequent aiding of others. Simultaneously, cheating would parasitize the altruists, causing them to lose fitness via the costs associated with altruism toward others without the compensatory benefit of later reciprocation. Of course, this difficulty might itself select for ability to identify cheaters and withhold altruism from them and/or to ostracize them (Trivers, 1971) which might in turn select for greater subtlety among the cheaters, and so on. Because of these problems, the practical significance of reciprocity as a mechanism for the evolution of altruistic behavior is not well understood at pres-

95

ent. However, sociobiologists seem to agree that, of all species, it is probably most likely to be of importance in *Homo sapiens*, a little known and rather aberrant primate.

Parental Manipulation

Brief mention should also be made of another theory for the evolution of altruism via individual selection, namely *parental manipulation* (Alexander, 1974). The question arises as to the generation level at which selection actually operates. In maximizing the fitness of individuals, does selection maximize the production of offspring, grandchildren, or succeeding generations? Specifically, when conflicts of fitness arise between offspring and their parents, who wins? The argument for parental manipulation is that parents will prevail in these circumstances, largely because offspring possessing a genetically mediated tendency to win such conflicts at the cost of their parents' fitness will themselves eventually be sabotaged by their own offspring, who will tend to inherit the same characteristics. According to this reasoning, parents should be selected for the ability to manipulate their offspring into behaving in altruistic ways that ultimately enhance the fitness of the parent, perhaps by benefitting siblings even at the cost of the altruist's inclusive fitness. During the coming years sociobiologists will doubtless give further critical attention, both empirical and theoretical, to the possibility of altruism through parental manipulation.

General Consideration

In the general case, any interaction between two individuals should have consequences for the fitness of each of them:

		RECIPIENT	
		GAINS	LOSES
INITIATOR	GAINS	mutualism	selfishness
	LOSES	altruism	spite

96

Selfishness and mutualism do not pose any particular difficulty for evolutionary biology. Of greater interest are altruism and spite, behaviors that are incongruous because the initiator appears to lose fitness, thereby raising the question as to how such tendencies can have evolved. Spite, as defined above, has yet to be documented in any natural system. Its occurrence would be a paradox for natural selection, and its absence may be construed as support for the evolutionary schema of sociobiology. Altruism, by contrast, is common and as this chapter has attempted to show, it can readily be explained by systems of kin selection and reciprocity. In this model, the initiator actually *gains* fitness, rather than loses, as shown above.

True altruism implies that the altruist's inclusive fitness is actually reduced. Such maladaptive behavior should only occur adventitiously if at all, and it should not evolve. The other proposed mechanisms for the evolution of altruistic behavior include group selection, kin selection, reciprocity, and parental manipulation, all of which are diagrammatically summarized in FIG. 4.4. Indeed, the term altruism is probably unfortunate, both because it implies a cognitive process that may well be uniquely human and because any of the above evolutionary mechanisms involve benefit to the altruist genes that necessarily exceed the cost; accordingly, they are all selfish in the long run. But the term is well established in the scientific literature and will probably endure.

In practice it is often difficult to identify whether a given behavior is altruistic or selfish and, even if altruism is indicated, real problems emerge in specifying whether the behavior has evolved through group selection, kin selection, reciprocity, or parental manipulation. This is one of the major difficulties in studying the sociobiology of altruistic behavior. For example, consider the phenomenon of helpers at the nest, described above. Helpers could actually be selfish, especially if their chances of successful breeding are low while they are young; accordingly, they may be maximizing their own personal fitness by saving their energy, avoiding potentially damaging competition, and waiting to breed when they are older. While waiting, they may simply be putting their time to good use by helping at

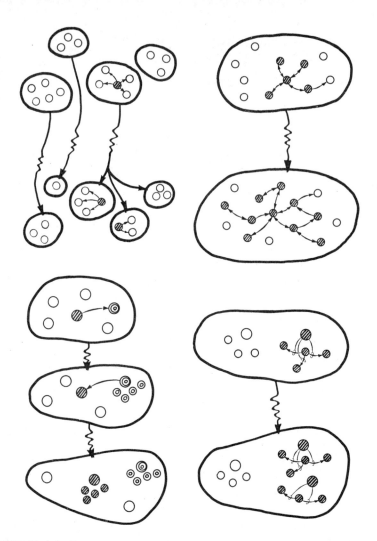

FIGURE 4.4. DIAGRAMS OF THE FOUR PROPOSED MECHANISMS FOR THE EVOLUTION OF ALTRUISTIC BEHAVIOR. (A) Group selection: the altruistic individual (striped circle) assists others within the group as a result of which the personal success of this individual within the group is diminished but that of the group is increased. (B) Kin selection: the altruistic individual assists its relatives, thus resulting in greater fitness of others carrying the same genes. (C) Reciprocity: the altruistic individual assists a nonrelative, thereby enhancing the latter's fitness, following which this recipient assists the original altruist, thereby ultimately enhancing its fitness as well. (D) Parental manipulation: a parent induces one or several offspring to assist its siblings at the cost of its own fitness but to the benefit of the parent, because the siblings receiving assistance are themselves more likely to reproduce successfully.(ILLUSTRATION BY D. COOK)

a nest and thus gaining child-rearing experience. Alternatively, this behavior may conceivably have evolved through group selection: Groups exhibiting reproductive restraint, because some potential breeders are helpers instead, may be less subject to injurious overpopulation and hence more successful than groups composed of all selfish breeders.

Or, helping could be selected through kin selection, in this case rearing siblings, as described. Furthermore, because kin often constitute groups, kin selection and group selection are often confounded and difficult to separate. Helping could also exemplify reciprocal altruism of a sort, if the helpers are buying adult tolerance in return for their child-rearing assistance. And finally, parental manipulation cannot be excluded in this case, because most helpers are in attendance at the nest of their own parents.

For another example and a rather gruesome one at that, consider the case of infanticide among langur monkeys in India. These animals live in harem troops controlled by a single adult male who does essentially all of the breeding. Nonbreeding males form peripheral, bachelor groups and attempt periodically to overthrow the harem master. When the previous patriarch is successfully ousted, the new dominant male proceeds to kill the infants in his newly conquered troop (Sugiyama, 1967). This behavior, although reprehensible by some human standards, represents a highly adaptive strategy for the adult male because he is not related to his victims. Furthermore, eliminating the infants not only reduces competition, it also has a more significant effect: while the adult females are nursing infants, these females do not ovulate; with their infants removed, these newly childless females begin to recycle, thus providing the newly ascendant male an opportunity to father his own offspring.

Similarly unpleasant but nonetheless adaptive strategies have been reported for African lions (Schaller, 1972) and, indeed, the slaughter of babes at the breast is a not-uncommon occurrence among human barbarians. It may seem surprising that bereaved langurs and lionesses reward their infants' murderers by copulating with them. But keep in mind that, in the "unsen-

99

timental calculus" of evolution (Wilson, 1975), differential reproduction is all that counts. Given a situation of childlessness, the best strategy (the most fit) is to breed again. To withhold copulations would be maladaptive spite; moral repugnance without compensating evolutionary advantages may well be a uniquely human luxury.

Alternatively, one possibility would be refusal to copulate with infant killers, if the threat of such refusal would induce males to accept nonrelated infants. Such a strategy could only work if it was adopted by all the females in a recently captured troop; it would be at the mercy of selfish cheaters who maximize their fitness by reproducing with the new male, regardless of his prior misdeeds. This strategy has apparently not evolved, indicating the difficulty of enforcing such collective societal sanctions when they conflict with individual self-interest.

Thus far, our account of langur infanticide has certainly not

PLATE 4.8 TWO ADULT FEMALE HANUMAN LANGURS (*at right*) CHARGING AN ADULT MALE WHO HAS JUST SEIZED AND SEVERELY WOUNDED AN INFANT. Such behavior by the females is altruistic in that it subjects the altruist to personal risk, for the benefit of another—the injured infant and its mother. (PHOTO BY S. HRDY/ANTHRO-PHOTO)

revealed any altruism, only extreme selfishness and, incidentally, no concern whatever for the good of the species. However, females clearly are disadvantaged by this grisly male strategy and, interestingly, they have been found to assist each other in defending their offspring, although with little success (Hrdy, 1974; Plate 4.8). Again, the basis for this altruistic female-female assistance cannot easily be identified. It could result from group selection, if despite the risk of being injured by the male while intervening on behalf of another female, the behavior is selected because of the ultimate benefit that befalls groups composed of individuals who behave in this manner. Alternatively, females may be assisting relatives and hence the behavior could be favored through kin selection. Reciprocity could also be involved: "I'll help you defend your offspring if you'll help me defend mine". And finally, parental manipulation cannot be excluded. Parents may be selected for the production of females who themselves tend to defend the offspring of their sisters, their nieces and nephews and, thus, the grandchildren of the original parents, despite personal fitness of the altruists being reduced in the process (Plate 4.9).

Regardless of the mechanism involved, it may well be significant that defenders of infants are most likely to be old females (Hrdy and Hrdy, in press). Such individuals most probably possess a low reproductive potential, that is, their altruism is purchased at low cost. It would be interesting to learn whether assistance is differentially directed toward infants whose reproductive potential is high and/or whose immediate danger is great but not hopeless—that is, high benefit to the recipient. For an intriguing sidelight, note also that adult females engage in an interesting strategy if they are pregnant at the time of the male takeover: They undergo a false estrus during which the new male copulates with them. This may well serve to deceive the male into tolerant acceptance of her offspring when it is eventually born, if he interprets it as his own—even though it arrives rather soon! The assumption is that male langurs are not very good at reckoning time although presumably, if the female strategy of pseudo-estrus were sufficiently frequent, it

PLATE 4.9 A GROUP OF HANUMAN LANGUR FEMALES AND INFANTS. Overlapping generations of close female relatives form the core of langur social organization. Considerable altruism occurs among these adult females, especially defense of infants against murderous males. (PHOTO BY S. HRDY/ANTHRO-PHOTO)

would select for male ability to discriminate their own offspring from those of their predecessor. The Chinese have a proverb: "It is a wise man who knows his own child".

For a final example showing the conflicting interpretations possible for animal altruism, consider the case of alarm calling, discussed earlier. At least among marmots for example, alarm callers do not seem to suffer mortality as a result of their behavior (Barash, 1975a), and alarm calling could actually be selfish. It could possibly enhance personal fitness by manipulating the behavior of others, whose flight makes them more apparent and thus susceptible to a predator (Charnov and Krebs, 1974). In addition to group selection, kin selection is also a possibility, as with reciprocity and parental manipulation. Indeed, the disentangling of these confounding biological bases for altruistic behavior constitutes a major challenge for future sociobiologists, bearing in mind that the various explanatory categories are not necessarily mutually exclusive.

Our fascination with biological interpretations of altruism may in part reflect our own conflictual state as humans. Religious, moral, and societal injunctions for human altruism may have been necessitated by the existence of biological tendencies for personal selfishness, tendencies that must somehow be overridden if complex human societies are to function smoothly in the modern world (Campbell, 1975). By contrast, animal societies are more consistently a reflection of biological imperatives; animal social systems may well represent neither more nor less than the sum of behaviors of individuals, each of whom is acting to maximize its inclusive fitness. Whether we admire the devoted altruism of worker honey bees or abhor their obligatory regimentation, we cannot deny the biology of their behavior, even as we ponder the roots of our own morality.

F I V E
The Social Question:
To Be or Not to Be?

The Darwinian dictum of survival of the fittest was interpreted
by many in the late nineteenth century as justifying the worst
excesses of laissez faire capitalism and its resulting social inequality. Thus, with dog-eat-dog seen as nature's way, it appeared unnatural and therefore somehow undesirable for
mercy, justice, and egalitarianism to intervene in the conduct of
human affairs. Actually, it is uncertain to what extent this misinterpretation of natural selection was an honest error or a de-
liberate fabrication by those with an interest in the prevailing
social system. In any case, among biologists the study of animal
social behavior became a neglected area and a problematic one
at that: It was difficult to reconcile the intensely selfish competi-
tiveness that was supposed to prevail in the natural order with
the rather smoothly functioning social networks that actually
existed. With every individual perceived as being concerned
only with its personal interests, cooperative social organization
was an anomaly indeed.

Productive study of animal sociality began with work of the
ecologist W. C. Allee (1938) and his students, who were at pains
to identify the adaptive significance of various animal grouping
patterns. Gradually, a picture has emerged that parallels our
conclusions regarding altruism: Animals appear to behave in
such a manner as to maximize their personal, inclusive fitness,
although such behavior may sometimes superficially appear to

involve a sacrifice in fitness. In the general case of social be-havior, there is now good evidence that insofar as individuals are social, they benefit by it; this is just as well because, if they didn't benefit, we would be hard pressed to explain why they are social. Actually, the evidence in this regard is still somewhat weak, since most studies of animal sociality have been evaluative only, although with the advent of sociobiologic thinking corre-lational approaches have also been attempted, and with increas-ing success.

Much of sociobiology's intellectual momentum derived from the startling insight that social behavior is amenable to evolu-tionary analysis. Accordingly, considerable attention has been directed to questions concerning the evolution and adaptive significance of social behavior, the main topic of this chapter.

Sociality

There are many advantages that animals derive from being so-cial, with different advantages accruing to different species, each associated with its unique pattern of social organi-zation. There are no across-the-board benefits of sociality per se; each species profits in its own particular set of ways, and ac-cordingly each demonstrates a somewhat different pattern. For example, all sexually reproducing species must engage in at least some minimal interaction between male and female, sufficient for fertilization to occur. Among most mammals, in-cluding nearly all of the rodents, the most abundant mamma-lian order, copulation itself provides the only adult sociality. The most persistent social bond is that between mother and young, which itself usually terminates at weaning (Eisenberg, 1966; Plate 5.1). Alternatively, male and female may form a dis-tinct social bond, monogamy, as in geese, songbirds, eagles, beaver, foxes, and gibbons. This seems to correlate, among other things, with advantage derived from assistance of two committed parents in rearing offspring and/or defending some valuable resource such as a territory (Plate 5.2).

PLATE 5.1 LEOPARD MOTHER WITH JUVENILE OFFSPRING. As with most mammals, these animals are rather asocial, with only a brief association between adult male and female, followed by a mother-infant bond that terminates when the young become independent. (PHOTO BY N. MYERS)

Animal mating systems vary enomously, largely as dictated by ecological constraints (see Chapter 6). Certainly, mating considerations are a primary determinant of social organization. In general, sociality in itself is advantageous in facilitating male-female encounters. Among relatively solitary animals, meeting and coordination between male and female may be difficult and time consuming; indeed, the Indonesian rhinocerous, a species that is currently endangered because of habitat destruction as well as over-hunting, is in particularly dire circumstances because it is essentially solitary, relying for reproduction upon chance encounters between male and female. With the population drastically reduced, such encounters have become rare, thus endangering the species yet further (Ripley, 1952). The blue whale may be in a similar boat (no pun intended).

PLATE 5.2 SCARLET IBIS PARENTS WITH THEIR TWO CHICKS. Both parents assist in rearing their offspring. (PHOTO BY NEW YORK ZOOLOGICAL SOCIETY)

But nothing is all good or all bad: Although sociality may facilitate male-female encounters, it may also reduce the likelihood of any one individual breeding successfully because of the increased competition among individuals for sexual partners. In such a system, males are more likely to be excluded reproductively, because a small number of males is more capable of inseminating a large number of females than females are of monopolizing male reproductive effort. As a result, we would expect males to show less affiliation with the social group than females. This is generally the case. Furthermore, asocial or dispersing males and/or those that leave the reproductive social unit, forming bachelor groups, tend to be subadult juveniles and/or subordinate adults, as predicted by evolutionary theory. These are the individuals that are most likely to be reproductively excluded, if they remained, and that are therefore most likely to enhance their fitness by seeking new reproductive opportunities in other social units (Plate 5.3).

There is abundant evidence that dispersing individuals usu-

PLATE 5.3 A GROUP OF FOUR BACHELOR MALE LIONS IN THE KALAHARI DESERT. In this sparse environment, prides are dominated by a single male who excludes other males. These reproductively excluded males band together and may occasionally attempt to take over a pride. (PHOTO BY R. WOLFF)

ally suffer a high mortality (Errington, 1963), but *reproductive* failure is often tantamount to mortality insofar as evolution is concerned. Furthermore, at least in some species such as Arctic ground squirrels (Carl, 1971), dispersers may balance the fitness loss implied by likely mortality with the occasional, though rare, chance of an enormous reproductive payoff, if they succeed in establishing a new group of their own. Furthermore, the high frequency of male dispersers may also be encouraged by aggression from the older males, who are defending their own reproductive interests. Of course, if the dispersers are eventually successful, this also enhances the fitness of the original parents. It would be interesting to know the extent to which the behavior of human subadults and of adults toward their children is consistent with these general trends. Certainly, adult-offspring antagonism seems to peak at adolescence and "going off to seek one's fortune" is not unique to nonhuman animals.

Beyond opportunities for mating, sociability in general has many other direct consequences for reproductive performance. Of these, care of the young ranks high (see Chapter 7). A social network comprised largely of kin provides numerous opportunities for altruistic assistance in the rearing of offspring (Plate 5.4). Sharing food, direct provisioning of young, helping to carry someone else's young (*aunting* in primates), and defense and warning against predators are all possible and, indeed, frequent. Furthermore, gregarious living makes possible the receipt of other forms of altruism among adults as well, mediated perhaps by group selection, kin selection, reciprocity, and even parental manipulation. Altruism could thus be a factor in the evolution of sociality, insofar as social individuals by virtue of their mutual altruism are more fit than asocial, selfish members of the same species who do not profit from the assistance of others. On the other hand, social organization may also permit the evolution of cheating and intraspecific deceit that may render sociality a less fit strategy than solitary life. In any event, as with most of the presumed adaptive values of animal sociality, it is not clear to what extent the supposed advantages of altruism are a legitimate *cause* of sociality rather than a *result;* that is,

109

once sociality has evolved for whatever reason, altruism may be potentiated, although it may not actually be responsible for the initiation of sociality. In one sense, however, this whole discussion may be moot: Both sociality and altruism exist, and they clearly are interconnected. This may be the most important point.

Other factors also undoubtedly receive heavy weighting in the evolutionary equation that ultimately produces the special pattern of social organization that characterizes each species. Consider predation: It is a strong selective force in shaping animal social organizations. As part of a social group, each individual is able to profit from the eyes, ears, and noses of all the other group members and therefore is probably more alert to potential predators than it would be if it had to rely entirely

PLATE 5.4 SOCIAL GROUP OF ELEPHANTS, INCLUDING ONE JUVENILE. The closely-integrated social system of elephants provides for considerable adult care of young animals, extending beyond the solicitude of the mother. (SAN DIEGO ZOO PHOTO)

upon its own senses. Accordingly, Olympic marmots feeding socially interrupt themselves to look around, presumably scanning for predators, significantly less often than do solitary feeders (Barash, 1973a).

To live gregariously is to become a fibre in a vast sentient web overspreading many acres; it is to become the possessor of faculties always awake, of eyes that see in all directions, of ears and nostrils that explore a broad belt of air; it is to become the occupier of every bit of vantage ground whence the approach of a lurking enemy might be overlooked. (Galton, 1871)

On the other hand, grouped prey may be a more conspicuous target than solitary individuals that could escape detection more easily. The disadvantage of conspicuousness has been suggested as a major factor selecting for dispersed nests as opposed to colonial breeding among many species of birds (Crook, 1965). However, cross-species generalizations relating grouping and predation risk may be risky, as evidenced by the fact that among some species, American bisons or African wildebeests, for example, the individuals are themselves so large as to render increased conspicuousness through grouping a negligible added risk. On the other hand, the advantages of increased watchfulness and sexual opportunities may be substantial (Plate 5.5).

In an article with the intriguing title "Geometry for the Selfish Herd," Hamilton (1971) proposed that animal grouping could develop if each individual behaves to increase the likelihood that its neighbor rather than itself will be taken by a predator. Thus, assuming that a predator is likely to take the closest prey that it encounters, the latter can increase its survival if it surrounds itself with susceptible neighbors. A randomly appearing predator will then be more likely to encounter one of these neighbors first. This model carries the interesting implication that animal sociality may in part be a selfish behavior that is ultimately adaptive for the individual but detrimental to the species.

Other suggestions relating grouping patterns to decreased likelihood of predation have involved the *confusion effect* that

PLATE 5.5 ADULT MALE BISON SURROUNDED BY FEMALES AND ONE CALF.
During the breeding season, males compete with each other for social dominance, which confers access to sexually mature females. Despite the prehistoric existence of prairie wolves and grizzly bears in North America, bison were not significantly endangered by their conspicuousness in herds, because of their large body size. (PHOTO BY D. LOTT)

may be generated by large numbers of prey, thus making it difficult for the predator to concentrate upon a single individual (Humphries and Driver, 1967) and the possible mimicry of a larger animal by an aggregation of smaller ones; this notion has been applied particularly to schools of fish. Mathematical treatments have related gregariousness to the sense capacities by which predators detect their prey, to the prey's response, whether concealment or evasion, and to the likelihood of detection combined with its consequences for the survival of individual group members (Triesman, 1975a, b).

The evolutionary influence of predators upon animal sociality goes beyond simple issues of detection; once detected, prey species are often more capable of defending themselves when grouped than when alone. Passive defense may occur, as when a loosely organized flock of starlings coalesces when approached by a peregrine falcon. This predator dives toward its

112

prey at incredible speeds, literally knocking a victim out of the air with its clenched talons. A solitary bird is thus a tempting target, but when such individuals form into a dense group, a diving falcon risks serious injury, because it may also strike an unintended animal with some vulnerable part of its own body (Mohr, 1960). Flocking is therefore an adaptive response among the prey of falcons.

Alternatively, there are numerous examples of active defense against predators in which the cooperative efforts of several individuals achieve a result that would be impossible for solitary animals. For example, the males of a baboon troop interpose themselves between the females and juveniles and their predators such as cheetahs or leopards. A phalanx of adult male baboons is a formidable defensive shield, and single individuals are rarely found on the predator-rich savannahs, at least, not for long! Many prey species, especially birds, commonly engage in mobbing behavior, wherein adults harry a predator that is often individually far superior to them, e.g., crows mobbing an owl, starlings mobbing a crow, and sparrows mobbing a starling, reminiscent of the observation, "Even fleas have smaller fleas to bite 'em; and so it goes, *ad infinitum!*" For an excellent account of the anti-predator strategy of a socially nesting species of gull, see Kruuk (1964).

Another presumed anti-predator phenomenon related to gregariousness is *sexual dimorphism*, overt differentiation between males and females, as in size, structure, or aggressiveness. Although dimorphism may well derive from sexual selection (see Chapter 6), it has also been implicated in defense against predators, especially among primates (Crook, 1972; Plate 5.6). In many ways, males are the more expendable sex: Once fertilization is achieved, fathers are considerably less important than mothers for the eventual success of their offspring. This is particularly true of mammals, in which females are uniquely adapted to provide nourishment for their infant offspring. Rephrased in the language of evolutionary biology: Under such circumstances and especially with predation constituting a substantial risk to their offspring, males can maximize their fitness by becoming proficient and eager body-

113

PLATE 5.6 MALE GELADA "BABOON" GROOMS A FEMALE. Sexual dimorphism is particularly striking among ground-dwelling primates, where it is due to selection favoring both defense against predators and success in competition among males for access to females. (SAN DIEGO ZOO PHOTO)

guards, whereas females are most fit when they are good mothers.

Furthermore, since a mature adult male may have fathered many offspring within the social unit, he will have a reproductive interest in the succcess of many individuals within that unit; he will be selected for defending virtually any infant within his troop, whereas females will be selected for restricting their defensive behaviors to their own offspring and recognized close kin. Male Alpine marmots are significantly more likely to give alarm calls than are females (Barash, 1976b). As mentioned previously, reproductive competition among males tends to be much greater than among females. Not only does this contribute additional pressures for the evolution of dimorphism (see Chapter 6), it may also be partly responsible for the greater intensity of courageous anti-predator behavior among males than among females. Thus, since a large evolutionary payoff of dominance and its mating prequisites accrues to a small number of males, risky strategies become worthwhile. The physical and behavioral characteristics that promote success against predators are also likely to be of value in male-male competition. And vice versa. In this sense, defense against predators may be seen as an essentially selfish male strategy, as with defense of offspring. On the other hand, altruistic anti-predator behaviors might also be selected among males whose future reproductive potential is low because of low social status or old age, i.e., a low risk to the altruist (Chapter 4).

This line of reasoning may also be pertinent to male-female differences in human physique and temperament as well. Thus, the proto-hominids ancestral to modern *Homo sapiens* are generally acknowledged to have evolved on the same predator-rich African savannahs that support modern-day baboons, with their closely organized anti-predator sociality, strong sexual dimorphism, and highly aggressive males. It is therefore conceivable that human sociality itself was selected in part for its adaptive significance as an anti-predator strategy, both against four-legged carnivores and possibly against other bands of primitive humans (Alexander, 1971).

God may not always be "on the side of the bigger battalions", but groups are almost always dominant over solitary individuals, and larger groups win out over smaller ones. Thus, insofar as the outcome of aggressive competition has important consequences for fitness, social grouping may be of adaptive significance because individuals participating in such groups may be reproductively more successful than are soloists. To be more precise, genes that predispose individuals toward a social life may be better represented in succeeding generations than genes for solitary existence, if grouped individuals are more successful (per individual) in competition for territories, scarce water resources, or calories derived from successful defense of a hotly contested kill. Sociality in and of itself may accordingly generate selective pressures favoring the maintenance and elaboration of further sociality, insofar as it may contribute to success in within- and/or between-species competition. For example, larger groups are generally able to dominate smaller groups and among our prehistoric human ancestors, grouping may well have been adaptive, because it conferred increased success in competition with large predators, such as lions and hyenas, as well as other groups of early "men".

Many small birds form distinct, social flocks in the winter. This has generally been attributed to predator avoidance and/or increased foraging efficiency (see Morse, 1970). However, competitive factors may also be involved, as suggested by the observations that among black-capped chickadees, solitary individuals interact aggressively with other solitaries and are dominated by flocked individuals (Barash, 1974c). Flocking may thus provide greater freedom of movement and/or greater energetic efficiency while moving; that is, fewer calories are wasted on aggressive interactions.

Not only is it important to animals that they minimize competition and avoid becoming food for someone else, they also must obtain their own food and, here again, sociality is doubtless of adaptive significance. Just as two heads are better than one in detecting predators, the same also applies to detecting potential food. Of course, there are no free rides in nature; it helps to have help finding food, but once it is found, it must also be shared (Plate 5.7). Dispersed food resources therefore

PLATE 5.7 THREE CHEETAHS AT A KILL. Clearly, one disadvantage of group foraging is the necessity to share food once it is obtained. On the other hand, to some extent this can be compensated by obtaining prey that is sufficiently large to ensure that there is enough to go around. (PHOTO BY R. EATON)

favor the evolution of dispersed (solitary) foraging, and clumped food selects for clumped (group) harvesting strategies as the most efficient techniques (Horn, 1968). Among species such as squirrels, in which individuals harvest food and bring it to a centrally located nest, efficiency is maximized by a relatively solitary existence, thus avoiding the duplication of effort that follows from two or more individuals covering the same ground (C. Smith, 1968).

Social organization may also provide the opportunity for exchange of information concerning profitable foraging locations, thus conferring possible advantages to roosting aggregations (Ward and Zahavi, 1973) as well as colonial breeding (Emlen and Demong, 1975). For example, individuals that have been less successful in searching for food could note the direction from which others with a full crop or stomach had returned. Alternatively, they could follow them the next morning. Social grouping could be adaptive as sources of information concerning foraging areas when food is localized, abundant enough to support several individuals, and shifting in location, such as insect swarms, fruiting trees, or schools of fish (Plate 5.8). In such cases, the "information center" aspect of sociality could be selected by any of the mechanisms previously described for the biology of altruism (Chapter 4), or it could be a simple example of individuals parasitizing the information possessed by others.

Although a social grouping often provides advantages in terms of food detection, in some cases there are corresponding disadvantages through warning the prey. Thus, the footsteps of certain foraging shorebirds literally signals their approach to the sand-dwelling marine invertebrates upon which they feed (Goss-Custard, 1970). There is a limit to how light-footed even

PLATE 5.8 DAYTIME ROOST OF INDONESIAN FRUIT BATS. (PHOTO BY R. TENAZA)

birds can be, and in this case gregariousness is apparently selected among these species *despite* its food-getting disadvantages because of compensating advantages relative to detection and evasion of *their* predators. As expected, gregariousness in the case of each species of shorebird represents a unique and optimal compromise between the conflicting demands of predator avoidance on the one hand and avoidance of maladaptive foraging interference on the other.

Many social animals experience another advantage in food acquisition; namely, they can cooperate in obtaining prey, thus enabling them to tap food resources that would not be available to solitary hunters. For example, group-hunting wolves (Mech, 1970), African hunting dogs (Estes and Goddard, 1967), and hyenas (Kruuk, 1972) are able to prey upon moose, zebras and wildebeest, respectively, animals that are more than a match for these predators, were they hunting alone (Plate 5.9). As expected, sociality of predators tends to correlate with the size of the prey relative to themselves. Wolf packs feed on moose, elk and caribou whereas solitary (or pair-bonded) foxes hunt rabbits and small rodents. It is unclear to what extent sociality predisposes toward large prey or vice versa. But it clearly is hard to share a mouse.

In general, herbivores do not derive food-gathering benefits per se from social cooperation that are comparable to those gained by predators (but see Cody, 1971). However synchrony of breeding may exemplify a form of selectively advantageous proto-cooperation, in which individuals reduce the probability of their offspring being lost to a predator by in effect glutting the predators with potential food during a restricted calving season (Estes, 1966). Thus, predators can eat only a certain amount of food per day and newborn animals are among the most vulnerable. If births are spread throughout the year, then each individual can be picked off in turn. In contrast, if all births occur during a short period, as they often do, then predators may eat heartily during this brief time, but the chances of any one prey animal losing its offspring are greatly reduced, because under these conditions predators will take only a small percentage of those available. Thus, in unity is strength, both

119

PLATE 5.9 CLAN OF HYENAS, BRINGING DOWN A WILDEBEEST. Despite their reputation as scavengers, hyenas are formidable hunters and in fact, lions often scavenge off hyena kills! Hyena hunting success is largely due to their social grouping, which enables them to kill prey several times their size. (PHOTO BY N. MYERS)

for predator and prey alike (except when it isn't!). Of course, bear in mind once again that strength here is ultimately of evolutionary significance only as it translates into fitness, which it almost certainly does in the above examples.

Breeding synchrony may be advantageous not only in surfeiting predators but also in providing mutual stimulation that enhances reproductive performance (see the review in Farner and King, 1971). The sights, sounds, and possibly smells of other courting individuals appears to facilitate and synchronize courtship (Hailman, 1967), with the enhanced courtship of each pair feeding back into the pool of stimulation available to the other, and so on. Large gull colonies produce more offspring per nest than do smaller colonies (Darling, 1938). This is the *Darling effect* and has been attributed to greater mutual stimulation in the larger social system. Alternatively, larger colonies are usually inhabited by older birds which are in any case likely to produce more successful offspring than are younger parents; the Darling effect is therefore confounded and remains unproven, although it is a viable hypothesis. Furthermore, peripheral individuals are likely to be more susceptible to predation (Tenaza, 1971), just as human frontier settlements were at greater risk than more central locations. This may contribute to higher mortality in smaller groups: Because of the geometry of small versus large groups, a higher percentage of individuals is at the exposed periphery in the former than in the latter.

Mutual reproductive stimulation between couples is actually a special case of a more general consequence of animal grouping, namely *social facilitation*. This refers to the observation that among many species grouped individuals often do more of something and/or do it more readily, for a longer time, etc. than do solitary individuals. For example, goldfish and chickens eat more when they are maintained socially rather than in solitary confinement. However, the important process here may not be social facilitation itself but rather its occasional absence. Thus, a variety of ecological factors, translated into evolutionary pressures, have doubtless selected for a social grouping of one sort or another and have made some form of sociality part of the normal environment of all species. Given this, it should only be expected that, if individuals are deprived of the opportunity to assort normally, they will show various behavioral pathologies. Sociality may accordingly be as important to normal biological function as adequate food or rest. Indeed, isolate-reared rhesus monkeys show grossly abnormal behaviors (Harlow and Harlow, 1962), a result that should surprise no one familiar with the intense social interactions characteristic of most primates.

Social organization is often beneficial to individuals through an effect generally describable as *biological conditioning of the environment*. Many species require a highly modified environment in order to be successful; some must change their surroundings drastically in order to survive at all. In most such cases, solitary individuals are unable to produce the effect that cooperating social units can easily achieve. For example, prairie dogs and marmots each take advantage of the extensive burrow systems dug by the numerous members of a colony (King, 1955; Barash, 1973a). They also help maintain an appropriate environment by nibbling brushy plants (prairie dogs) and tree seedlings (marmots), thus preventing an ecological succession that would otherwise transform their grassy homeland into a less suitable habitat. Social insects such as bees and termites create micro-environments within their hives, often maintaining them at temperature and humidity regimes that differ dramatically from conditions outside. Many species can detoxify environmental poisons; they are more effective at this when grouped than when solitary.

On the debit side, sociality also creates an increased risk of environmental contamination as well as local overpopulation with all of its attendant disadvantages, including habitat destruction and increased likelihood of transmission of disease. Perhaps more than any other species, humans require the type of environmental modification that only social cooperation can provide. More than any other species, we are also capable of disastrous environmental manipulation, the consequences of which are becoming increasingly apparent.

When animals live in groups, the opportunity exists for division of labor, with all of the attendant efficiencies rediscovered in this century by Henry Ford. As might be expected, division of labor reaches its maximum among the social insects, correlating with their evolution of altruistic sterile castes. Thus, unless one specializes in reproduction itself, as have the insect queens that are basically egg-laying machines, extreme concentration upon one function occurs at the cost of overall fitness and therefore would be selected against. Among the peculiarly altruistic social insects, however, we find such curiosities as enormous soldiers whose mouthparts are so specialized for warfare that they cannot feed themselves and must be fed by other specialized workers; other "nasute" soldiers that are walking squirt guns, generating a stream of irritating or sticky fluids;

"repletes" that serve as living storage tanks of nutrient fluids; and members of certain termite castes, the geometric shape of whose heads precisely match the entrance to their hive where they obediently stand as living doors, moving aside only when they receive an appropriate knock from a colleague.

A fascinating scheme has been devised for the evaluation of adaptive function in insect division of labor, taking into account the relative fitnesses of hives containing different proportions of individuals specialized for different tasks (Wilson, 1968). The resulting system of *caste ergonomics* describes an optimal system that can evolve only through group or extreme kin selection, because it is based upon adaptive value for the group rather than the participating individuals.

Vertebrates are extreme individualists by contrast, as expected among organisms for which individual selection is paramount. However, some division of labor does occur, as among the social carnivores in which some individuals stay with the young while others participate in the hunt. Helpers at the nest exemplify a similar division of labor. However, in all these cases, defined roles are not characteristic of an individual for its entire life, as occurs among many invertebrates. The reproductively selfish vertebrate societies are composed of individuals that may temporarily assume particular duties, but all remain generalists insofar as they retain the capacity to reproduce and an ardor for doing so. All this is understandable to anyone familiar with the imperatives of evolution.

Perhaps the closest thing to division of labor among vertebrates is sexual dimorphism, mentioned earlier with regard to defense against predators. In addition to its possible derivation via sexual selection, sexual dimorphism could also have evolved because of an advantage in minimizing ecological competition among mated pairs (Selander, 1966). Different environmental demands by males and females would increase the fitness of both over *monomorphic* pairs, among whom physical overlap produces greater competition and, hence, lower fitness. In any event, humans are extreme if not unique among vertebrates in our division of labor via exchange of goods and services, i.e., butchers, bakers, and candlestick makers.

A final advantage of sociality is its potential for learning and the passage of tradition. Insofar as learning is achieved by trial and error, or other simple processes, a solitary animal can do as well as a social one. But other forms of learning are also possible. Thus, many animals acquire information by observational or imitative learning, and such cases obviously cannot occur unless individuals are in proximity. Such learning enables the propogation of beneficial social traditions. For example, European titmice (birds, not mice) learned to follow milkmen on their morning deliveries and then to open the lids and drink the cream before most householders rescued the remaining, freshly skimmed milk. This nefarious technique originated within a small population and rapidly spread to much of England (Fisher and Hinde, 1948).

The snow monkeys of Japan, a species of macaques, are provided with food by researchers who study their social behavior in otherwise seminatural surroundings. Among the commonly distributed foods are sweet potatoes and wheat. One time a particular adult female discovered that she could rinse the potatoes in sea water, thus cleaning them and also apparently conferring a delicious, light salting. This new tradition of potato washing spread rapidly through her troop. Sometime later, this same brilliant individual also discovered that she could separate wheat from sand by dropping handfuls of sandy wheat into water. Again this behavior soon became characteristic of her troop (Kawai, 1965). It is also interesting to note that the initiator's social inferiors copied her behavior most readily; those above her in the dominance hierarchy were more conservative and reluctant to adopt such radical departures from normal behavior. Again it is tempting to extrapolate to human behavior: Who are the vanguard of new ideas, the young or the old? The disinherited or the established? In any case, it also pays to recall that such a pattern is almost certainly adaptive: Who has the most to lose if the behavior proves to be maladaptive?

One reason why rats have been so difficult to eradicate, despite intensive trapping and poisoning campaigns, is their capacity for social transmission of such adaptive information as the existence of poisoned baits. Thus, if a rat has a bad experience with a poison or observes another animal that does, the rat will avoid that substance in the future, and this behavior will be observed by others who in turn will behave similarly. Incidentally, this explains the value of slow-acting rat poisons that cannot be associated with their effects. Rat poisons, potato washing, and milk-bottle opening all reflect the occurrence of socially mediated traditions, none of which would be available to solitary animals.

Certainly most human wisdom is culturally transmitted across generations, between groups, and from innovators within groups to the other members of the group. The capacity for rapid transmission and assimilation of new and adaptive behaviors was almost certainly favored in human evolution, perhaps especially during those long years on the savannahs when our brain size was increasing rapidly, thus suggesting strong selective pressures favoring behavioral flexibility and innovation. This notion provides for an interesting conceptual bridge between biological and cultural factors in the evolution of human social behavior. Social learning and the passage of traditions are clearly cultural, while the *capacity* to perform such operations are biological in terms of both the neural competence of individuals and the tendencies for social organization that make these operations possible.

By this time it should be clear that there is no unitary adaptive significance to social behavior. Rather, nature provides a kaleidoscopic array of advantages and disadvantages, depending upon the characteristics of the species itself, its *phylogenetic inertia* (Wilson, 1975), and its ecological situation. From this complex matrix, each species extracts its own optimum pattern of sociality. There have been numerous attempts to classify animal social organization. Because the basic orientation of sociobiology is evolutionary, the best classification for our purposes should emphasize function (Table 5.1).

In evaluating any social pattern, we must also bear in mind the danger of misjudging a specie's true social organization because of our own perceptual biases as humans. For example racoons and red foxes are generally considered typical solitary

TABLE 5.1
A Proposed Classification of Animal Grouping Patterns
A Species May Form Different Kinds of Groups at Different Times (*Modified from Brown, 1975*)

Type of Grouping	Definition	Examples
1. Aggregations	Groups formed by simultaneous attraction to a common source, rather than to each other, and/or the result of physical factors acting on individuals.	Earthworms under a rock; Bears at a garbage dump; Gulls following a plow; Hawks migrating along a ridge; Human spectators at a sporting event.
2. Survival groups	Groups formed largely by non-breeding individuals based on mutual attraction; the members are only randomly related.	Herring schools; Wildebeest and zebra herds; Night roosts of starlings; Winter foraging flocks of blackbirds.
3. Mating groups	Groups within which breeding occurs; this may involve a monogamous pair, a harem, or a troop (several males and several females), and often includes extended families, formed by offspring remaining with the parents rather than dispersing.	Gibbon and bald eagle pairs; Elk harems; Baboon troops; Mexican jay extended families; Human families.

TABLE 5.1 (*cont'd*)

Type of Grooming	Definition	Examples
4. Colonial groups	Groups formed by distinct breeding pairs or harems, often unrelated to each other, who seek the close company of other, similar breeding units.	Gull and other seabird colonies; African weaverbirds; Some bats and seals; Most human residential systems.
5. Unisexual groups	Groups of one sex, usually male, formed because of the breeding system: either as a means of attracting mates (lekking groups) or because of failure to do so (bachelor groups).	Prairie chicken or sage grouse lekking groups; Uganda kob lekking groups (a species of antelope); Bachelor groups in elk deer and seals; Human fraternities and sororities?
6. Clonal groups	Clones; groups formed by asexual reproduction and therefore composed of individuals each of whom are genetically indentical to each other and are typically in permanent physical contact.	Certain marine invertebrates, such as jellyfish polyps; Multicellular organisms?

carnivores; this because we, as visually oriented animals, only rarely *see* them associated together. However, they apparently are social in that neighbors recognize and respond differentially to each other as opposed to animals separated by greater distances. This pattern of social organization is probably maintained by chemicals, *pheromones,* contained in urine, glandular secretions, etc. These chemicals are deposited on trees, bushes, or rocks that are visited in turn by neighboring animals (Barash, 1975b).

Communication

Regardless of the reasons for sociality and the precise form that it takes, its occurrence makes certain demands upon the participants, or, alternatively, there are certain prerequisites for the evolution of sociality. This may be another case of which came first, the chicken or the egg. Sociality requires sufficient sensory development for each individual to be responsive to another's presence and also requires sufficient neural and locomotor capacity to orient one's behavior with regard to other conspecifics. Sociality also correlates with communication, the capacity for influencing the behavior of another through one's own behavior. A detailed analysis of animal communication is beyond the scope of this book. Rather, two points will be considered briefly: the relation of social pattern to communication and the evolution of communication.

Intensity of communication does not necessarily vary in any consistent way with the degree of social integration. A dog and a snake may interact only once, but the communication that passes between them, a mutual threat, may nevertheless be very intense. Social integration within a species *does* correlate with the subtlety and complexity of communication that occurs, although once again it is probably meaningless to worry about cause or effect. A good example comes from comparative studies of canids, viz., wolves, dogs, foxes, and coyotes. Foxes are characterized by simple, relatively asocial behavior patterns (Plate 5.10). They show a more limited communicatory reper-

PLATE 5.10 TWO YOUNG KIT FOXES (about two months old) DEMONSTRATING
SIMPLE AND EXAGGERATED COMMUNICATION. The animal on the right is showing defensive threat, with back arched, ears back, tail down and head low,
whereas the animal on the left is showing a "dominant" approach, with stiff
gait, ears up, body in "forward" position and tail high (not visible in this
photo). (PHOTO BY M. BEKOFF)

toire than do wolves, whose complex social organization is
paralleled by great range and subtlety of communication between individuals, using facial features, body postures, and vocalizations as well as presumed chemical communication (Fox,
1970).

Just as social behavior has evolved, so has communication.
Animals are capable of understanding each other to a remarkable extent, often with little or no prior experience. Genetically
mediated tendencies exist for behaving in particular ways in
particular circumstances and for responding appropriately to
the behavior of others. In other words, evolution can operate
on communication and has done so. Infant rhesus monkeys,
removed from their mothers at birth and shown photographic
slides, respond to the threatening face of an adult male by *presenting*, getting on all fours and elevating their posterior, a sign
of submission in most primates (Sackett, 1966). Bear in mind
that these infants had never before seen another monkey, let
alone a threatening one; they had never experienced the af-

termath of threat, i.e., an attack. Furthermore they developed the appropriate response rather rapidly, at about the age when young animals are increasingly independent of their mothers and increasingly likely to encounter less tolerant adult males. This is *ritualization*, the evolutionary modification of behavior to serve communication (Huxley, 1923), and it can be explained by natural selection operating upon the behavior of individuals. Consider that animal A performs Behavior 1 (opening mouth and staring at animal B) that functionally precedes Behavior 2 (animal A biting animal B). It would accordingly be advantageous (increase its fitness) for animal B to understand Behavior 1 and respond to it in some appropriate manner, so as to avoid Behavior 2. Individuals who did this would probably produce more successful offspring than would other individuals who were constantly getting bitten because of their obtuseness. In other words, selection should favor the identification of Behavior 1 as a threat. Of course, communication requires a sender as well as a receiver, and this poses no evolutionary problem. The threatening animal would also be more fit if it could get its way without actually biting its opponent, especially since the potential victim may be a relative. Furthermore, even a smaller and weaker animal could conceivably inflict injury during a fight. Selection would therefore also favor exaggeration of Behavior 1, prolonging the stare, exposing the teeth, vocalizing loudly, etc., thus promoting evolution of both the communicative behavior, the threat, and the response to it, in this case presenting.

The study of animal communication is a new and vigorous field (cf. Sebeok, 1968; Hinde, 1972; Otte, 1974), and both learned and genetic factors are often involved. Recent efforts to identify the evolutionary background of human communication by use of an historical approach have been provocative if not altogether convincing (van Hooff, 1972). In any event it seems evident that at least the *capacity* for communication is a genetically mediated human characteristic. We cannot communicate like bats, but they cannot speak like us. In fact, even chimpanzees can't quite do the trick, although they come astonishlingly close (Premack, 1971; Gardner and Gardner, 1971). We possess the most complex and varied social organization of any animal and communciation abilities that are comparably distinguished. Would anyone call this a coincidence?

Some Correlations

The social system characterizing a species also feeds back upon the early experiences of individuals with others of their kind, their pattern of *primary socialization*. Among members of the deer and elk family, social organization is fundamentally matrilineal, organized around adult females and their descendants. The primary social bond during childhood is accordingly between fawn or calf and its mother (Altmann, 1960). In contrast, among pack-living canids, peer-group association is predominant. Not surprisingly, primary socialization in these species occurs among littermates (Scott and Fuller, 1965).

Conceivably there may be another chicken-and-egg issue here: whether behavioral ontogeny or the social pattern that correlates with it can be identified as the motive force of this closely integrated developmental system. However, in this case social organization is probably the end toward which evolution is aiming, with ontogeny the proximate mechanism of achieving that end, rather than vice versa. This seems reasonable, both because of the great precision of adjustment between social systems and ecological situation, and because of the apparent ease with which primary socialization could be modified to suit the appropriate end product. Of course, this is not to deny all developmental constraints upon social organization. A full appreciation of sociality ought to consider the role of early experience in shaping the behavior of group members, just as the study of developmental psychology and sociology would profit by considering the nature of the social system(s) in the service of which various developmental processes have undoubtedly evolved.

As discussed earlier, most sociobiologic considerations of animal social organization have been evaluative. However, some correlational attempts have also been made, and a few of these will now be described. Basically, such research into the ecology of social systems represents simply a different way of slicing the sociobiologic cake: comparing the social systems of related species in different environments and correlating behavioral differences with the environmental differences, as opposed to starting with ecological factors such as predator avoidance and then looking for examples of how these factors influence sociality across a diversity of species.

The red-winged and tricolored blackbirds are two closely related species that are physically very similar. However, their social systems differ substantially: Red-winged males establish and defend relatively large territories (2,500 to 32,000 ft^2) within which they mate with one or more females. They defend their territories for many weeks and do much of their foraging there. In contrast, tricoloreds are much less evenly dispersed: They travel and breed in huge flocks, establish very small territories (35 ft^2), and defend them only briefly. Feeding occurs off the territory, and the young are provisioned by both the male and the female, as opposed to the red-wings, in which the male has virtually no child-care role, being heavily occupied with territorial defense and efforts to secure additional females.

These differences in social pattern are all attributable to a basic environmental difference: Red-wings are adapted for homogeneously spaced, temporally predictable, and persistent food sources. In contrast, tricoloreds have specialized in the exploitation of highly concentrated but unpredictable and often short-lived foods such as locusts that appear in outbreaks and the occasional vegetational plenty associated, at least before human settlement, with flooding in the central California river systems (Orians, 1961). Thus, the social system of tricolored blackbirds follows an opportunistic get-rich-quick strategy, while red-wings pursue a more conservative policy associated with access to established resources. A third species, yellow-headed blackbirds, occupies environments like those of the red-wings; as expected, its social system is also of the red-wing type.

The sociobiology of African ungulates, the large, hoofed mammals such as antelope and buffaloes, is now increasingly understood (Estes, 1974; Jarman, 1974). Forest dwellers tend to be relatively solitary whereas plains dwellers are highly aggregated. This correlates with the physical problems of maintaining group cohesion in broken habitats, abundance of predators, and nutritional requirements. The food needs of each species vary particularly with its body size, such that smaller animals must specialize on foods of high caloric density. These are generally shrubby plants, that are themselves unevenly distributed compared to the lower-quality but more homogeneously dispersed grasses. Migration patterns, feeding style, anti-predator response, body size, and to some extent degree of sexual dimorphism can all be tied together in this analysis.

In North America, mountain sheep form coherent, integrated social units whereas moose are essentially solitary. This correlates with the dispersal of young among the latter (Altmann, 1960) and its absence in the former (Geist, 1972), which in turn is explicable in terms of their different environments. Sheep occupy stable, climax environments such that most suitable habitats are generally filled and optimum strategy is to remain with established and successful groups and acquire, for example by social tradition, information concerning home range and migration patterns. In contrast, moose are nonmigratory; they inhabit relatively transient, subclimax environments, and young animals can discover a new and suitable habitat only by leaving the old.

Understandably, primates have long fascinated us. They have also provided some of the most exciting possibilities for sociobiologic analysis as well as some of the most frustrating near misses. An early attempt to classify primate societies into evolutionary grades revealed certain apparent correlations with habitat and diet (Crook and Gartlan, 1966). Solitary species include many of the primitive prosimians (lemurs, galagos, and tarsiers) that also tend to be nocturnal, insectivorous and forest dwelling. Small family groups such as gibbons are diurnal and more vegetarian but still generally forest dwelling, whereas occupation of savannahs correlates with larger size of group and omnivorous habits. But the patas monkeys of Africa live in grassland and assort in small family groups, whereas squirrel

133

monkeys of South America are forest dwellers and often travel in enormous bands.

Subsequently, a more satisfying framework has been proposed, based upon the degree of male integration into the social unit (Eisenberg, Muckenhirn, and Rudran, 1972). The following categories were recognized: solitary; bi-parental family; single-male harems, one male and several females; age-graded male troops, adult males tolerant of subadult males, as in gorillas; and multimale troops, two or more fully adult males, as in macaques and chimpanzees. Unfortunately, there remain many exceptions and inconsistencies in the analysis of primate sociobiology, especially the enigmatic social parameters of many forest-dwelling primates, which are difficult to study and often refuse to behave as predicted (Clutton-Brock, 1974). The human penchant for placing observations into neat, exclusive categories may have to accept some disappointments in this area for some time to come. Nonetheless, for valuable compilations of recent work in this area, see the following: Jay (1968), Crook (1970b), Kummer (1971), Dohlinow (1974), Michael and Crook (1974), Teleki (1974), and Tuttle (1975).

Actually, the failure of many primate species to conform to a neat theoretical plan should not be surprising, as it exemplifies another, perhaps neater, generalization: Primates are highly intelligent, behaviorally flexible creatures, and we are only beginning to appreciate the complexity of environmental factors relevant to each species, as well as the range of potential responses of each species and the adaptive consequences of these responses. For example, East African baboons have been found to vary their social patterns as a function of whether they inhabit savannahs or forests (Rowell, 1969) and "urban" rhesus monkeys in India are socially distinct from their "rural" counterparts (Sinh, 1969). Modifiability of this sort emphasizes the role of experience in adjusting the social behavior of big-brained animals, but at the same time it does not negate the role of evolution: The ability to vary social patterning in an adaptive manner as a function of environmental conditions is *itself* adaptive and has undoubtedly been selected. The limits of possible social organization are established by genotype, as is the range of actual environment-society correlations that characterize each situation.

Finally, although social organization is most pronounced among members of the same species, associations between species often exist. Baboons and impalas often occur together and are sensitive to each other's alarm calls. Individuals of both species profit from the association, because baboons have good vision and impalas have good hearing and sense of smell. Many insects, especially among the ants, are dependent upon other species for their survival, the systems ranging from slave making (dulosis) to utter parasitic dependence (inquilinism). One of the most fascinating examples of interspecific associations doesn't exemplify sociality to any great extent, but it is a fascinating example of the precision of behavioral adjustment that can be achieved by evolution, even when it operates *between* species.

A fairly common interspecific behavior among birds is *nest parasitism*, in which one species deposits its eggs in the nests of another, relying upon the deceived foster parent to rear the offspring. As expected, this generally results in an evolutionary race between parasite and host, with the victimized foster parents selected for ability to discriminate their own eggs from those of the nest parasite and the would-be child abandoners selected for the production of eggs that mimic the host's own. A species of South American cowbirds lays its eggs in the nests of oreopendolas and caciques, members of the blackbird family. These two host species themselves suffer a high mortality from botflies, dangerous parasites that destroy nestlings. Now, when oreopendolas and caciques nest near certain wasps or bees, these insects somehow repel the botflies, thus reducing nestling mortality and enhancing the birds' reproductive success. Without protection conferred by wasps or bees, oreopendolas and caciques are victimized by the botflies *unless* they are parasitized by cowbirds.

Nests of caciques and oreopendolas containing cowbird young are actually more successful than are nests that have not been parasitized, so long as no wasps or bees are around, because the young cowbirds remove the botfly eggs and maggots not only from themselves but also from their foster siblings. So, oreopendolas and caciques that are nesting near wasps or bees discriminate against cowbirds; they are already protected from botflies and will lose fitness if their own nestlings lose some food to the young cowbirds from which they do not otherwise benefit. On the other hand, oreopendolas and caciques nesting away from wasps or bees do not discriminate against the parasitic cowbirds; they are more fit with them than without them. The arrangement is yet more precise: There are two types of cowbirds, those laying mimetic (resembling the host's eggs) and nonmimetic eggs. Predictably, the nonmimetic form specializes in host species living away from wasps and bees; these hosts welcome the parasite, thus increasing their own fitness, so there is no selective advantage to deceiving the host as to the nature of the additional eggs. And the mimetic form lays eggs in host nests that are protected from botflies; these hosts discriminate against the cowbirds who are therefore selected for the production of eggs that cannot be distinguished from those of the host. The published paper first describing this intricately counterbalanced system was appropriately titled "The Advantage of Being Parasitized" (N. Smith, 1968).

S I X
Strategies of
Mate Selection
and Reproduction

All living things are strategists. *The Random House Dictionary* defines strategy as a "plan, method or series of maneuvers for obtaining a specific goal or result". Of course, animals do not consciously engage in strategic actions with particular outcomes in mind, or in any case we need not assume that they do. Rather, given that individuals have been selected for behaviors that maximize the ultimate evolutionary success of genes influencing those behaviors, organisms should behave *as though* they were seeking a specific goal or result, in this case maximum inclusive fitness. Conscious volition is irrelevant here. Imagine two alleles, A_1 and A_2, competing for a particular *locus* (place) on a chromosome of a given species: If A_1 produces a behavior pattern that makes its possesser more successful than does allele A_2, then A_1 will eventually replace A_2 in the population, and the behavioral strategy exemplified by A_1 will have been selected.

Organisms are faced with numerous strategic choices during their lifetimes: how to avoid predators, where to find food, what to do when the seasons change and the environment becomes stressful and, conversely, how to respond when the environment becomes so conducive to life that competitors abound. These and many other issues present numerous opportunities for analysis. However, because reproduction is so closely related to evolutionary success, we might expect that reproductive strategies would be particularly susceptible to natural selection. Indeed, the analysis of reproductive strategies has been one of the most productive areas of sociobiology (Barash, 1976c) and this and the following chapter will attempt to outline the major reproductive decisions that all living things must make.

How to Reproduce? Sex or No?

The disadvantages of sex are obvious; its compensations are less apparent. A sexually reproducing animal must often devote substantial time and energy to attracting a mate. Time spent singing courtship songs, for example, is not available for building a nest, finding food, or just plain resting. The acquisition of a mate often requires making oneself relatively conspicuous through loud sounds, gaudy appearance, and/or eye-catching behavior, thus making the individual more conspicuous to predators as well. If a species is rare, getting the two sexes together may itself be a considerable problem. Even when a species is abundant, individuals of one sex may still go unmated if the ratio of adult males to females is unequal. For example, if the reproductive unit is monogamous, one adult male and one adult female, and there are more adult males than females, then some males must forego reproduction.

On the other hand, some individuals of one sex or the other may go unmated even if the adult sex ratio is equal, if the reproductive unit is either polygynous, one male mated to several females in which case some males are excluded, or polyandrous, one female mated to several males in which case some females are excluded. In addition, efforts to secure a mate may involve the ardent suitor in personally damaging competition with others of the same sex. Finally, even after a mate is acquired, there is the danger of desertion or cuckoldry. For reasons to be discussed later, the former is a particular problem for females; the latter, for males. Anyway you view it, sex is clearly a hassle.

In addition to these obvious difficulties, sex imposes a subtle but no less profound evolutionary cost. A sexually reproducing organism shares 50% of its genes with any one of its offspring, the genes contained in either its sperm or its egg. In contrast, the organism would have 100% of its genes represented in its offspring if it reproduced asexually. Thus, sexual breeders appear to suffer a 50% loss in fitness relative to their asexual counterparts (Maynard Smith, 1971; Williams, 1975; for an alternative view, see Barash, 1976d). Because asexual breeding produces individuals that are genetically identical, we would also expect such offspring to benefit their parents by showing maximal altruism toward each other, thus resulting in smoothly functioning social cooperation. Indeed, this is what we find, for example, in the asexually reproducing coral polyps (Wilson, 1975). In fact, every multicellular animal owes its coherence to lack of competition among its constituent cells that in turn can be attributed, at least in part, to the genetic identity of these cells. Each cell of a human, for example, has been produced by the asexual reproduction of a fertilized egg. In contrast, sexual reproduction produces genetically distinct offspring, thus introducing *self* versus *other* into the biological world. By generating individuality and hence opening the door to selfish competition between individuals whose personal strategies often conflict, sexuality would seem to constitute a further hazard to those would-be parents who adopt this particular strategy of reproduction.

However despite all these difficulties, sex is very much with us. Nearly all vertebrates reproduce sexually. The advantages of producing genetically distinct progeny most likely outweigh all the disadvantages described. Many organisms with the opportunity to fill up an otherwise empty environment will often reproduce asexually, such as daphnia (water fleas) in a mud puddle or aphids on a newly colonized sunflower. But sexual reproduction and dispersal generally follow deterioration of the environment. Sexual breeding reshuffles the genetic cards of the offspring produced and thus results in a better investment for the evolutionary future. Again, bear in mind that this investment need not be assessed directly in any way by the participating individuals. Rather, this assessment was itself conducted by the evolutionary process: Those parents that

139

adopted sexuality as a reproductive strategy under particular circumstances left more offspring than those employing a different strategy. G. C. Williams (1975) has provided a careful exposition of the ways in which sexual strategies can exceed asexuality in maximizing the fitness of its practitioners. Basically, sexual parents hedge their bets against future environmental change by generating a sufficient number of varied offspring to give them an advantage over practitioners of the more conservative, asexual strategy.

When to Reproduce?

Since most complex animals have opted for sexual reproduction, we shall assume it in considering the other strategic decisions required by a reproducing animal. The question of when to reproduce can be considered in two ways: At what developmental stage in an individual's life should reproduction be initiated and under what environmental circumstances? Unlike the yes-no decision of sexual versus asexual reproduction, these and other strategic decisions considered here do not involve a simple choice of distinct alternatives. Instead, they require decisions on a graded or sliding scale. But even though the choice represents some point on a continuum, we can assume that the ultimate determining criterion is the same as for the simpler, yes-no case: Animals adopt the strategy that maximizes their fitness.

It would seem advantageous for animals to breed when they are as young as possible, because an early start would maximize the number of offspring for which each individual is responsible. Indeed, this is mathematically correct, and in the absence of other confounding factors selection would push reproduction to earlier and earlier ages. This is in fact what happens, and selection has resulted in the earliest possible age at reproduction, *consistent with the production of a maximum number of surviving offspring and relatives.*

There are several costs associated with breeding at too young an age. If reproduction occurs before dispersal, there is the

140

danger of parent-offspring or sibling mating. This inbreeding results in decreased fitness of the offspring produced and hence a decrease in the fitness of those parents who mate before dispersal. This would select for delayed maturation, earlier dispersal, and/or an ability to distinguish relatives from nonrelatives and to refrain from mating with the former.

Beyond this, there is good evidence that reproduction constitutes a physiological stress resulting in higher mortality. It therefore would be good strategy to delay breeding until the animal is sufficiently large and strong that it can optimally withstand the strain; in the long run an individual may do more to enhance its fitness by starting later, if this means the eventual production of more and healthier offspring. Healthier parents would almost certainly produce more successful offspring, and an increased lifespan of the parents would also contribute to this final result.

The physical strain of reproducion is almost invariably greater upon females than upon males; this is particularly true of birds and mammals in which females must generate large eggs or a placenta, nutrition for the unborn infant(s) and in the case of mammals, milk after birth. It might therefore be expected that within the same species, females would be selected for reproducing at a later age than males. However, this does not seem to hold, presumably because the advantages of early reproduction among females exceed the physiological costs. In fact, in many birds and mammals, particularly the polygynous species, the opposite holds: Males generally become sexually mature later than females. This happens, for example, in seals (Bartholomew, 1952, 1970) and in grouse (Wiley, 1974). This surprising observation is a consequence of the sociality of reproduction in such animals. Thus, when a small number of males inseminates a large number of females, competition among the males is intense, just as the evolutionary payoff is great. Cause and effect cannot be distinguished clearly in this case. If individuals engaging prematurely in sexual competition lose out to larger, more experienced competitors in addition to suffering a reduction in their likelihood of future reproductive success, then *sexual bimaturism* will be selected; members of one

141

sex will mature later than the other. Many species are characterized by vigorous battles among the competing males, so this result is not surprising (Plate 6.1). Optimal strategy for a young male in such a species is to defer reproductive competition until it is old enough so that its size and experience renders success more likely, for example in mountain sheep (Geist, 1971), red-winged blackbirds (Peek, 1971) and elephant seals (LeBoeuf, 1974). Similar analysis may also suggest an ultimate explanation for the sexual bimaturism of our own species—boys mature later than girls.

If greater age correlates with greater experience and size, which in turn results in greater potential reproductive success, we can expect members of one sex to mate preferentially with those members of the other sex that offer them greater reproductive success. Older males, for example, may well control resources that enhance a female's reproductive opportunities;

PLATE 6.1. TWO ELEPHANT SEAL BULLS THREATEN EACH OTHER. Titanic battles often occur among these males and not surprisingly, sexual maturity among males of this species is not reached until four years of age, whereas females (*in foreground*) mature earlier. (PHOTO BY B. LE BOEUF)

this may involve a better territory or certain personal qualities such as hunting or defensive abilities. Females would then be most fit if they discriminated against younger males and males might therefore enhance their own fitness if they "bided their time," all the while gathering the experience needed for eventual reproductive success. In some cases, delayed reproduction may be good strategy for both parents, as in European oyster catchers, shore birds that require several years of learning to become sufficiently adroit at opening their mollusc prey so as to be able to raise a family.

A simpler strategic decision concerns when to breed in terms of the appropriate environmental circumstances, the optimum breeding season. The most consistently relevant factor here seems to be available food for the young, and a good generalization is that animals will mate so as to produce young whose maximum food demands are made when food is most readily available. In the northern hemisphere, food is generally most available in spring and/or summer, so optimal strategy for ungulates, deer and their allies, is to mate in the autumn and give birth in the spring. On the other hand, the lag between mating and the production of young is much shorter in birds, so for them breeding typically occurs in the spring or early summer. Animals living in environments less predictable by seasons cue their reproductive activities to other features that are consistently correlated with enviromental factors that ultimately lead to reproductive success. Desert amphibians commence courting in response to sudden and erratic rains, and African weaver birds are cued to the appearance of green grass (Ward, 1965) such that seeds are available by the time the young are in the nests. Once again, these adaptive correlations are not fortuitous: During the evolutionary history of each species, the reproductive strategy of each component individual was tested by natural selection. Successful strategists, those that were better adapted, left more offspring than the unsuccessful, so that each species tends to be composed of successful strategists. Desert toads that entrusted their moisture-loving eggs to rain pools have left more offspring than those whose genetic makeup led them to deposit eggs indifferently upon the dry sand.

With Whom to Reproduce?

A sexually reproducing animal must select an appropriate mate. This is the major function of *courtship*, a characteristic set of behaviors occurring between male and female directly concerned with mating and reproduction. Courtship often tends to be elaborate, conspicuous, and stereotyped. It is often quite consistent among individuals of the same species and distinct between different species. This suggests a substantial genetic component to courtship, and indeed this appears to be true in a wide range of species: *Drosophila* (Spiess, 1970), fish (Clark et al., 1954), and ducks (Sharpe and Johnsgard, 1965). The reason for this seems fairly clear: Choice of a mate has great consequences for an animal's reproductive success. The cost of errors may be high and the payoff for a good choice is often very great. This is especially true of species that reproduce only once during their lifetime as opposed to those that breed several times. Considering the importance of making a good decision when there are no second chances, one can predict that species of this type would invest their courtship with a greater degree of fail-safe genetic programming than would the latter.

But in what sense can a courting animal make a mistake? One of the simplest and most serious potential errors is attempting to reproduce with a member of the wrong species. The resulting hybrid offspring, if any are produced at all, are nearly always less fit than are offspring produced by matings between members of the same species.* This is not surprising. Each species is a very precisely functioning set of genes, a complex of DNA in which the male and female contributions are co-adapted to each other (Mayr, 1970). Genes from different

---

*These *interspecific* hybrids, offspring produced by mating between individuals of different species, should be distinguished from *intraspecific* hybrids, offspring produced by mating of individuals that are of recognizably different genetic background but still of the same species. These latter individuals may actually be more fit, demonstrating the hybrid vigor so popular with animal breeders, while the former tend to be evolutionary failures.

144

species are not mutually co-adapted, and therefore the outcome of such a mixing may be less fit in a variety of ways: less viable, less fertile, or abnormal at any stage from fertilization of the egg, or the failure of fertilization, up to maturity of the offspring (Dobzhansky, 1970).

Imagine animals A and B, both of species X, both inhabiting an environment also occupied by species Y. Animal A is more discriminating in terms of the courtship signals to which it will respond. It will mate only with other members of species X and either rejects or is insensitive to the overtures of species Y. In contrast, animal B is less finicky and mates with a member of species Y. The interspecific hybrid offspring of animal B is relatively weak, courts ineffectively, is sterile, or perhaps does not even survive. Clearly, animal B's lack of mating discrimination has placed it at a great disadvantage compared to animal A that mated appropriately and produced normal, healthy offspring. Animal A, the discriminating one, will leave more offspring than will animal B, and insofar as the tendency to make a discriminating choice of a mate is under some genetic influence, genes favoring care in selecting one's mate will predominate in the population. Simultaneous with the ability to discriminate, i.e., receive and act upon signals from an appropriate mate, natural selection will also generate the ability to transmit easily identified signals to members of the opposite sex. In this manner, natural selection will favor the evolution of genetically mediated tendencies to mate preferentially with members of the correct species. This is the use of courtship as an *isolating mechanism*. For a further discussion of isolating mechanisms, see Brown (1975, Chapter 18).

Ethologists traditionally identify three major functions of animal courtship: advertisement, overcoming aggression, and achieving coordination, both behavioral and physiological, between mates (Etkin, 1964). Both sexes must agree on whether to copulate and when; in many cases, further cooperation is necessary in order to defend a territory, provide adequately for offspring, etc. (Plate 6.2). Courtship helps assure coordinated behaviors of this sort and significantly, courtship is most prolonged in species with extensive cooperation among mates. The

PLATE 6.2 A MATED PAIR OF HERRING GULLS "CHOKING" TOGETHER AT THEIR BREEDING COLONY (*center of photo*). This mutual courtship activity serves to reinforce their agreement as to the nesting site and is also used as an aggressive display in coordinated defense of breeding territory. (PHOTO BY J. GALUSHA)

advertisement function of courtship can be divided into several components: attracting attention, identifying species, identifying sex, and identifying reproductive condition. For example, the familiar springtime song of the red-winged blackbird male says the following to a prospective mate: "Here I am; I am a red-winged blackbird; I am a male; and I am ready for sex." In most vertebrate animals, the role of advertiser is performed by the male, while the female is the discriminating customer. Males tend to be selected for salesmanship; females, for sales resistance (Williams, 1966b). The reason for this distinction goes to the heart of male-female differences and is crucial to an understanding of the sociobiology of courtship and mate selection.

It is easy enough to tell a man from a woman or even a male giraffe from a female giraffe. But how about sparrows or warblers or indeed virtually all birds in which there is no differ-

146

ence in external genitalia, although the male is generally more brightly colored than the female? Indeed, why do we designate the former as male rather than the other? What of oysters, where there may be no apparent external differences and yet we still speak confidently of males and females? The underlying biological meaning of male and female derives simply from the nature of the reproductive cells, the *gametes*, produced. One sex specializes in the production of a small number of relatively large gametes, eggs. We call these individuals female. The other sex produces a large number of relatively small gametes, sperm. These are the males. The evolutionary reason for this dichotomy is unclear. For a thought-provoking theoretical treatment of this difficult area, see Parker et alia (1972).

Regardless of the causes of the male-female distinction, the consequences are profound. The sex that produces a small number of large gametes (female) has much more invested in each of those eggs than the other sex has in his sperm, and this asymmetry of interest is particularly apparent in higher vertebrates. Among birds, for example, the eggs laid by a female may constitute upwards of 25% of her total body weight (Welty, 1963), and among mammals the small initial size of each egg is more than compensated for by the large investment required in generating a placenta, nourishing the embryos from her own bloodstream, running all the risks associated with pregnancy and childbirth, and, finally, producing milk after the offspring are born. In contrast, male birds and mammals produce sperm in incredible abundance with each ejaculation, and these sperm can be readily replaced, usually within a very short time. Males can usually walk or fly or move by some other means away from the consequences of copulation. For females, the issue is more serious. The consequences of a bad decision thus fall particularly heavily upon females and hence it is not surprising that they tend to be the more discriminating sex. At minimum, males invest a little bit of time and sperm; females may have to live for weeks, months, or a lifetime with the consequences of their reproductive decisions. Small wonder that they are fussy.

It should therefore occasion no surprise that a walk through a pond in spring may commonly reveal bullfrog males clinging

147

to the toes of one's boot in the characteristic posture of *amplexus*, used to clasp a female as part of mating. Male bullfrogs will clasp almost any appropriately sized object. The female, on the other hand, is more discriminating: She will release her precious store of eggs only if stimulated in just the right way, often by sound as well as touch, and by a male of the right species. The males are following a strategy that is most adaptive for them—play fast and loose. The best female strategy is different. With this fundamental contrast in optimum male-female mating strategies in mind, it should now be clear why males are usually the aggressive sexual advertisers while females are usually the careful comparison shoppers. The few exceptions to this rule are particularly instructive and will be discussed in the following section.

Traditional ethological wisdom has it that, in addition to advertising and coordination, courtship also serves to reduce aggression between reproductive partners. Particularly among certain carnivorous invertebrates where male and female often differ greatly in size, with the female generally the larger, elaborate male courtship seems to increase the likelihood of his winding up a mate rather than a meal. Among vertebrates, male territory owners are often highly aggressive and frequently inclined to respond aggressively to any intruder upon their domain, whether male or female. Such cases of apparently mistaken identity are especially common among species in which males and females are not dramatically different in appearance. In situations of this sort, courtship often involves the display of certain physical characteristics that distinguish males from females.

For example, among the North American woodpeckers known as flickers, males have a black line on each side of their face, whereas females do not. These birds apparently distinguish male from female by the presence or absence of this trait, since when researchers painted a pair of "mustaches" on a nesting female, she was attacked by her mate (Noble, 1936). Alternatively, aggression may be reduced and the sexes identified by characteristic behaviors that distinguish the individual unequivocally as either male or female. In some cases, as with the

North American song sparrow, female courtship mainly involves tolerating the male's aggression, without provoking it any further, until he becomes habituated to her presence. In other cases, characteristic postures, vocalizations and/or chemical secretions may be employed.

In addition to these presumed adaptive advantages of courtship, sociobiology suggests a new area of potentially fruitful inquiry relating courtships and strategies of mate selection. If we consider the maximization of successful reproduction to be a concern of individuals rather than partnerships, it becomes clear that strategies may be unique to each individual, depending upon its sex and also varying as that individual varies in age, experience, future reproductive potential, the number of offspring and relatives with which to interact, and so forth. The ultimate selfishness of natural selection causes each individual to make the best possible decision, for *its own* maximum fitness, and not necessarily that of a prospective partner. One of the most important reproductive decisions an animal makes concerns its choice of a mate, and courtship may therefore serve a primary role in permitting each participant to assess its potential partner. This is the notion of courtship as evaluation.

Individuals could enhance their fitness by evaluating a potential mate in several respects. Among many species, mates rely upon the physical prowess in the partner, as in defending a territory, warding off predators, or obtaining elusive and/or dangerous prey. In such cases, it would benefit each individual to ascertain the physical competence of a potential mate before committing its resources, whether eggs, nest or den site, or simply its precious time and energy during the breeding season. This may explain the elaborate aerial courtship of many falcons and eagles, for example, during which objects are exchanged in mid-air. Any lack of speed or coordination would be apparent during these demanding acrobatics. These animals depend upon the hunting ability of their partner in providing food for their young, and it would therefore be maladaptive to mate an incompetent partner. Selection would accordingly favor courtship that permitted individuals to make a decision which enhanced their reproductive success: If your mate is inadequate, better to know it sooner than later. It has even been proposed that secondary sexual characteristics may constitute disabilities that themselves reduce the ecological fitness of their carriers (Zahavi, 1975). Under this controversial view, characters such as bright and conspicuous plumage or large and energetically expensive antlers serve as markers for individuals of the opposite sex by indicating the underlying competence of any individual capable of flourishing despite the disability.

Assessment may also occur more directly. Behavioral inclinations to defend a territory or mob a predator, for example, may also be assessed during courtship. Many species of ducks engage in *inciting* during courtship, in which the female appears to induce her betrothed male to attack a strange male (Plate 6.3). This behavior has been considered a classic example of ritualization. In this particular case, inciting may also help the female to assess the aggressiveness of a prospective mate. To demonstrate that this is so, it would be necessary to show that females reject males who respond inappropriately and that some correlation occurs between preferred levels of male aggressiveness and reproductive success.*

Among those species in which the committed assistance of both mates is required in order to rear offspring successfully, it may be in the interests of each to assess whether the other is available for full-time parental duties, unimpeded by a prior commitment to another individual. And insofar as the female makes a greater reproductive investment than does the male, such an assessment would be of particular importance to the female. Such an assessment could be achieved by prolonged courtship, since a lengthy betrothal increases the likelihood of discovering a third party, if one exists (Trivers, 1972). In fact, this strategy would be particularly successful since the mate of a would-be two-timer would also profit by keeping close tabs on its partner, thus increasing the likelihood of discovery. This may help to explain why members of monogamous species,

---

*In this regard, bear in mind that more aggressive does not always equal better. It is possible for hyperaggressive adults to experience reduced fitness if they devote excessive energy to fighting and insufficient attention to parental duties; this is *aggressive neglect* as described by Ripley (1961).

PLATE 6.3 FEMALE MALLARD (*at left*) INCITING HER MATE (*at right*) TO ATTACK A SECOND MALE (*foreground*). Such behavior on the part of the female is a ritualized part of courtship in mallards. (PHOTO BY P. JOHNSGARD)

those in which a single male and female make a reproductive commitment to each other, typically engage in lengthy courtship prior to actual mating. In contrast, species which are notable for the brevity of their courtship are generally those in which one partner contributes only its gametes to the reproductive success of the other. In these cases, the subsequent behavior of its mate does not influence the fitness of the one rearing the offspring, and accordingly, selection has not produced behaviors to assess such tendencies.

Among species requiring considerable parental care in the rearing of offspring, mechanisms should evolve that prevent individuals from "wasting" their time and energy on the rearing of unrelated offspring. For females, this problem is not especially severe, except for cases of nest parasitism (see the final pages of Chapter 5). Males, on the other hand, are more susceptible to this trap, because they cannot be completely confident that they are the fathers of the young they help rear. Therefore they should not only resist being culkolded, but they

151

should also refuse to pair with a female that shows signs of already having mated. Among ring doves, females engage in a prominent "nest-soliciting" display following exposure to a male. In a laboratory experiment, male ring doves were presented with two types of females: some that had previously been exposed to males and were "nest-soliciting" accordingly, and some that had not. The males exhibited less courtship and more aggression toward the former females than toward the latter (Erickson and Zenone, 1976). In other words, they behaved in accordance with sociobiologic theory in that they were disinclined to mate with females who had already been involved with other males. There may be parallels here in human behavior, whether reflecting cultural practices or evolutionary wisdom. In any case, this example emphasizes the basic "selfishness" of evolutionary biology applied to mate selection.

Because individuals may differ as to their optimum strategies, with each selected to follow that which maximizes its own inclusive fitness, there are numerous possibilities for the evolution of *intraspecific deceit,* defined here as the transmission of false information to a member of the same species. Misinformation as to one's mating status is only one possibility. Another is exaggeration of those characteristics being assessed by the mate.

In his book *The Descent of Man and Selection in Relation to Sex* (1871), Charles Darwin considered why males and females of the same species often differ so greatly from each other. He identified two processes as responsible for this distinction, processes that we now consider *intrasexual selection* and *epigamic selection* (Huxley, 1938). Although Darwin felt that these processes operated independently of natural selection, we now recognize that both these forms of *sexual selection* are achieved by differential reproduction of individuals possessing certain characteristics and hence are not truly distinguishable from natural selection itself. Intrasexual selection operates by competition between members of the same sex, usually the males, often resulting in large size, aggressiveness, and the presence of organs of threat and combat such as antlers, horns, tusks, manes, and canine teeth (Plate 6.4). Epigamic selection occurs

PLATE 6.4 STELLAR SEA LION BULL *(center)* SURROUNDED BY HIS HAREM OF FEMALES AND THEIR OFFSPRING. Intra-sexual competition among males results in selection for aggressiveness and large size, since those males possessing such traits are more successful in competition with other males and accordingly are able to win a harem and achieve reproductive success. (PHOTO BY B. LE BOEUF)

by the differential attractiveness of individuals of one sex to individuals of the other. This may lead to the elaboration of the same traits as found for intrasexual selection as well as extremes of sexual display such as the tail feathers of peacocks or lyre birds and the incredible antics of birds of paradise (Plate 6.5). In certain cases, epigamic sexual selection operates on behavior alone, with virtually no effect upon the physical structure of the animals involved. Thus, the bower-birds of Australia and New Guinea create definite structures (bowers) to which females are attracted. These bowers vary from simple to elaborate, and an inverse correlation exists between the gaudiness of bowerbird males and the elaborateness of their characteristic bowers (Gilliard, 1969). A drab suitor needs an elaborate bower; a well-adorned suitor need not concern himself with such a structure (Plate 6.6).

PLATE 6.5 WHITE PEACOCK DISPLAYING. An extreme example of epigamic selection—selection resulting from female choice of males possessing elaborate display characteristics. (SAN DIEGO ZOO PHOTO)

It is occasionally claimed that sexual selection operates to the detriment of a species and that in some cases it can lead to extinction. If members of a species become bizarre and highly specialized, they may well be less able to adjust to a changed environment, and extinction will follow. But selection itself could never lead a species over an evolutionary cliff. If large antlers or tail feathers threaten to become so disadvantageous as to reduce the reproductive success of their bearers, then the epigamically selec*ting* sex, usually the females, would be selected for preferring mates that are somewhat less adorned. Those that choose to mate with partners who are disadvantageously overdeveloped will produce offspring, one-half of which (the males) will suffer the same continuing disadvantage and hence will bring less evolutionary return than those who select mates whose characteristics lead to greater ultimate reproductive success.

PLATE 6.6 SATIN BOWERBIRD CONSTRUCTING HIS BOWER. These relatively plain-looking birds build elaborate bowers as part of courtship. (SAN DIEGO ZOO PHOTO)

Thus, like other forms of natural selection, sexual selection contains an inherent self-regulating feature. One might also expect that both intrasexual and epigamic selection provide numerous opportunities for the evolution of deceit, and in fact to some extent the elaboration of display characters itself can be seen as orchestrated deceit. But selection's feedback system operates here as well, with excessive deceit selecting for increased ability to distinguish the deceiver.

The sex investing more in the production of offspring, usually the female, will be a crucial resource for the sex investing less. In most cases, therefore, males will compete among themselves for access to females. This competition will enhance the social inequity in reproductive success. Thus, nearly all females will be mated, whereas among males, a minority will do most of the breeding, with other males going unmated. This is particularly true because males, by virtue of their small initial reproductive investment, can usually inseminate several females, and the large evolutionary payoff this represents will itself produce intense competition for females. In the general case, individuals of the sex investing less compete among themselves for access to individuals of the sex investing more. It is therefore generally true that males experience a greater variance in reproductive success than do females. In other words, in terms of reproductive success, disparity between the "haves" and the "have nots" is usually greater among males than among females.

Trivers (1972) has contributed a valuable clarification with the notion of *parental investment*, defined as any behavior toward offspring that "increases the chances of the offspring's survival at the cost of the parent's ability to invest in future offspring". As described earlier, a discrepancy generally occurs between the sexes with regard to the parental investment made in each offspring produced, with females usually investing more than males. With each offspring produced, total female parental investment will tend to rise more rapidly than will male parental investment, because the relatively large parental investment represented by each offspring detracts more substantially from a female's remaining reproductive potential than it does from a male's. However, reproductive *success* will increase equally for both sexes because it is always equivalent to the number of offspring produced (FIG. 6.1). The parent with the greater investment per offspring will be selected to cease reproducing sooner than will the other, because maximum net reproductive success of the former will be achieved with the production of fewer offspring. As a consequence, the sex investing less, generally the males, will seek access to additional individuals of the sex investing more, generally the females. The result is greater variance in reproductive success, greater intrasexual competi-

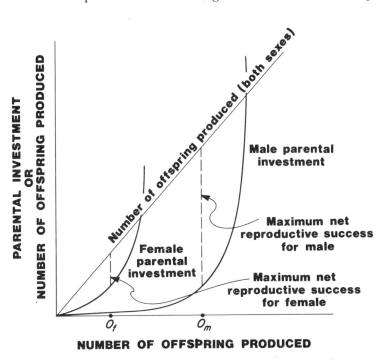

NUMBER OF OFFSPRING PRODUCED

FIGURE 6.1. PARENTAL INVESTMENT AND REPRODUCTIVE SUCCESS. Both males and females are selected for producing the number of offspring that results in maximum reproductive success. This is not measured simply by the number of offspring produced but rather, the *difference* between number of offspring produced (benefits) and decreased reproductive potential in the future because of the present offspring (costs). In the case shown here, it is assumed that female parental investment increases more rapidly with each succeeding offspring than does male parental investment. Since each sex is selected for maximizing the difference between the straight line and the parental investment (cost) curves, males and females will disagree as to the optimum number of offspring: members of each sex are selected for achieving the longest possible dotted line (net reproductive success; benefits minus costs). Accordingly, males are most fit when they produce O_m offspring, whereas females are most fit producing O_f offspring. Therefore, males will compete with other males for access to additional females (polygyny). In cases where the parental investment functions are reversed, females compete for additional males (polyandry). (*Modified slightly from Trivers, 1972.*)

157

tion, and more elaborate displays among individuals of the sex investing less (Plate 6.7).

Trivers' theory that sexual selection results from differences in parental investment can be tested by examining those cases where the usual investment pattern is reversed, the species in which the male makes a greater parental investment than the female. In such cases, males typically have the sole responsibility for parental care, thus making up for their small initial investment of sperm. The results accord with prediction: the females have apparently been subjected to sexual selection. They are larger, more aggressive, more brightly colored, etc. Among these are seahorses, certain frogs, and several species of marsh and shore birds (jacanas and phalaropes).Parental investment is an enormously valuable heuristic concept and one of the important cornerstones of sociobiology. In addition to its relevance for sexual selection, it generates important insights into parental behavior as well (see Chapter 7).

Biologists have generally concentrated their attention upon the behavior of the "sexually-selected sex": these individuals are more dramatic in both appearance and behavior, as befits their greater competition. But the other sex, generally the females, are more than passive recipients of male attention. They are the object of competition because they invest so much in reproduction, and the fact that they are competed for gives them the opportunity of mating preferentially with males that contribute maximally to *their own* reproductive success. In addition to the various possibilities for evaluation discussed above, females can also choose males that are most likely to provide them with food, for example, thereby enhancing their ability to produce healthy offspring. Females can accordingly select for males who themselves provide a large parental investment, even though such investment may not be strictly required by the reproductive biology of the species concerned. This process may have contributed to male provisioning of females during courtship (see Plate 3.1) and in some insects, it may have actually selected for male willingness to be eaten by the female after copulation! (Thornhill, 1976).

Differences in male-female reproductive strategies provide

158

PLATE 6.7 MALE THOMPSON'S GAZELLE *(right)* USES FORELEG KICK TOWARD FEMALE AS PART OF COURTSHIP IN THIS SPECIES. In nearly all vertebrates, males compete for access to females, and are the sexual aggressors. (PHOTO BY F. WALTHER)

other interesting possibilities. For example, a scientific paper was recently published with the eye-catching title "Prostitution Behavior in a Tropical Hummingbird" (Wolf, 1975). Hummingbirds forage especially at flowers and they compete among themselves for possession of feeding territories that include flowers, a preferred source of food. Because the males are more aggressive, they usually end up with most of the available flowers, forcing the females to make do on other foods. Outside the breeding season, males respond aggressively to any other hummingbird that intrudes upon their hard-fought, flower-rich territories. But in at least one species, males tolerate the presence of females who are even permitted to feed from their precious flowers . . . *provided* the females copulate with them! This behavior is something of a paradox insofar as male fitness is concerned, since no offspring are produced by these matings;

159

possibly, however, they help cement relationships leading to successful pairing later in the year. On the other hand, the issue is clearer for the females: they are engaging in a low-cost behavior that enhances their fitness by giving them access to resources controlled by males, that would otherwise be unavailable. The underlying mechanism may be different, but certainly the *facts* are closely analogous to human prostitution.

In What Social System to Reproduce?

Animals demonstrate a wide range of mating systems; a simple classification is shown in Table 6.1.

The actual designation of mating system type is sometimes arbitrary, as with many small mammals in which the male occupies a home range or territory that overlaps that of several females (see Chapter 9). The females apparently mate with this male, but it is unclear whether a bond of any sort is formed. Such a system may therefore be either promiscuous or simultaneously polygynous, depending upon how one chooses to define pair-bond. The terminology itself is not important. What is important is that each species tends to show a characteristic mating system—identification of the adaptive value of each system constitutes a major challenge to modern sociobiology. If species A has a characteristic anatomical feature, a particular bill shape for example, it seems reasonable to expect that feature to be functional in the life of the animal possessing it; i.e., we can investigate its adaptive significance, the evolutionary rationale for its occurrence. Similarly, with mating systems.

Because of the essential differences between the sexes in parental investment (described above), polygyny and promiscuity would seem to be the mating systems most preferred by males and they are, in fact, the most common social systems of mammals. Monogamy, where it occurs, appears to be a secondarily derived condition. However it is the most common system for birds and is practiced by over 90% of the known species (Lack, 1968). Monogamy usually correlates with near equality in parental investment by both mates. This suggests one major

160

TABLE 6.1
CLASSIFICATION OF ANIMAL MATING SYSTEMS

SYSTEM	DEFINITION	EXAMPLE
Monogamy	Reproductive unit of one male and one female; pair-bond formed	
Annual monogamy	Bonds formed anew each year	Small passerine birds: sparrows, warblers, etc.
Perennial monogamy	Bond retained for life	Swans, geese, eagles, gibbons
Polygamy	Reproductive unit of one individual of one sex and several of another; pair-bond formed	
Polygyny	One male bonded with several females	
Serial	One male bonded with several females during a breeding season, but only one at a time	Pied flycatchers
Simultaneous	One male bonded with several females simultaneously during the breeding season	Red-winged blackbirds, fur seals, elk
Polyandry	One female bonded with several males	
Serial	One female bonded with several males during a breeding season, but only one at a time	Rheas
Simultaneous	One female bonded with several males simultaneously during the breeding season	Jacanas
Promiscuity	No bonds formed	Grouse, bears, wildebeest

reason for its existence: Monogamy will be selected when successful reproduction requires the cooperation of two committed adults. Birds have a high metabolic rate and among species in which the young are helpless at birth, the voracious appetites of these offspring may require assistance from both the mother and the father. Because female mammals are uniquely adapted

161

to nourish their young with milk, males are limited in terms of how they can increase their fitness through direct interaction with their offspring (Barash, 1975c). As predicted (Orians, 1969) monogamy is accordingly rare among mammals, especially among herbivores, in which the low caloric value of their food makes provisioning by the male especially inefficient. Significantly, monogamy among mammals is most common among carnivores such as foxes and coyotes in which the male can share prey efficiently with the offspring.

Monogamy may also serve in facilitating defense of a scarce and valuable resource (Wilson, 1975). Certainly, defense could also be achieved by the cooperation of additional adults, but if the resource in question is scarce as well as valuable, the disadvantages of sharing could select for a small reproductive unit. A prolonged period of infant dependence, including the need for extensive learning by the offspring, could also select for monogamy.

It would seem that females should generally be most fit if mated monogamously, so long as all their available eggs are being fertilized. They profit reproductively from the assistance of a committed male. Males, on the other hand, might prefer polygyny, because of their low parental investment. On the other hand, if they spread themselves too thinly across several females and are therefore unable to invest sufficiently in their own offspring, they would be less fit than those males that mated fewer females but provided the parental follow-through that would lead to greater reproductive success. However, when monogamy is not indicated by the particular circumstances of a species, we can expect males to be more fit when polygynously mated. They should therefore seek to mate with more than one female; indeed, with as many females as possible, so long as each additional mate increases their Darwinian fitness. Insofar as some males are successful in their polygynous endeavors, either simultaneous or serial, this should also produce some unmated, bachelor males. Given that females would presumably profit from the undivided attention of such otherwise uncommitted males, why do females of many species agree to polygyny? Why should they mate with an already mated male when bachelors are available?

Several factors are relevant here. If, because of the nature of the species, males cannot contribute anything beyond their sperm to the reproductive success of their mates, then females would lose nothing by mating with an already mated male (polygyny), so long as he produces sufficient sperm to fertilize all her eggs. In addition, females would gain by mating with the best males around, if those males would help produce geneticaly healthy offspring that would themselves be more likely to experience reproductive success, hence increasing the fitness of females who adopt the strategy of mating with the best males. This would contribute to female acquiesence in harem formation, as in elk or fur seals. By successfully acquiring a harem, a male not only has indicated his physical competence, but he also is likely to possess characters appropriate to success in sexual selection, both interasexual and epigamic selection. Thus, his male offspring are likely to possess these characters, and a female who mates with him is therefore likely to experience greater fitness, via her grandchildren, than will a female who accepts a male who is unable to become a harem master.

Alternatively, females may be influenced more by the quality of the resources controlled by a male than by the male himself (Orians, 1969). The relevant environmental resource often is food availability or security of nest sites from predation. All things being equal, it is assumed that females are nonetheless most fit if mated monogamously. However, if the environment is heterogeneous and male A is controlling some resource that would contribute to the reproductive success of a female, she may maximize her fitness by mating with him, even though in so doing she foregoes paternal assistance that she could receive from an otherwise unmated male, B. For this to occur, the parental contribution of the males must be relatively small, and the difference in resource quality as it affects reproductive success must be relatively great, exceeding the *polygyny threshold* for each species (FIG. 6.2).

The various predictions suggested by this formulation have

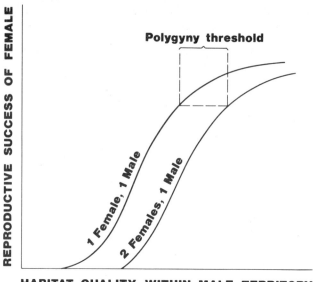

FIGURE 6.2. THE POLYGYNY THRESHOLD MODEL OF
HABITAT QUALITY. Female reproductive success increases as
the quality of her male's territory increases. Note that for any
given territory, females are more fit if they mate monogamously than bigamously. However, if the territories maintained by two males differs by a certain minimum amount—
the polygyny threshold—then a female is more fit if she reproduces in this better habitat, even though doing so requires
that she share her mate with another female. (*Modified slightly
from Orians, 1969.*)

been supported thus far. For example, polygyny should be
more common in heterogeneous environments, and it is. For
example, of the fifteen species of polygynous North American
songbirds, eight nest in marshes and five nest in grassland or
savannah habitats. Both marshes and grasslands are notori-
ously patchy in terms of resource distribution. Most songbirds
are monogamous; they tend to be forest-dwellers, in which male
and female cooperate in provisioning the young. As predicted,
polygyny is favored among precocial animals such as pheasants,
in which the relative independence of the young reduces the
need for paternal assistance. In a further test, the mating system of indigo buntings was evaluated (Carey and Nolan, 1976).

164

These small, woodland birds occupy diverse habitats which appear to vary greatly in quality from the territory of one male to that of another. The males are brightly colored, an apparent consequence of sexual selection. They spend about 30% of the day singing, much more than most passerines. Polygyny was therefore predicted for the indigo bunting: It was found that 10% of the adult males had two mates, while some had none. Despite the reduced male assistance, females mated polygynously were able to produce as many offspring as females mated monogamously. This, presumably, because they were on relatively better territories.

Polyandry can be derived from the same polygyny threshold model simply by endowing the females with control over patchily distributed resources, little maternal role, etc. However as expected from the discrepancy in male-female parental investment, polyandry is rare and the reasons for its occurrence are not well understood. In many species of birds the male commences incubation after the female has laid the eggs, while the female forages and recoups the metabolic losses she incurred in producing the eggs. In those situations in which she is able to gain sufficient nourishment to lay another clutch, it will be to her advantage to do so. In some species, the female then incubates a second clutch herself, fathered by her original mate (Pitelka et al., 1974). However, if she can induce another male to incubate this additional clutch, thus liberating the female for further reproduction, then so much the better for her. Of course, it would be maladaptive for a male to invest in offspring in which he has no genetic interest, so it would behoove the female to permit a different male to father each successive clutch, thus increasing the likelihood of his providing additional investment. For a further account of possible evolutionary routes to polyandry, see Jenni (1974).

Promiscuity, the final mating system to be considered here, is characterized by the absence of any pair-bond between the mates. However, this does not imply a lack of discrimination in selecting a mate. Indeed, promiscuous species often demonstrate extreme elaboration of secondary sexual characters, thus indicating a high level of sexual selection. Perhaps the most striking examples of promiscuous mating systems are found among the various species that form *leks*. Leks are specialized territories, maintained by individuals of the more competitive sex, usually the male, in close proximity to each other. Such territories are generally small and function entirely for mating. Females visit the lek, mate with one or a small number of the males present, and then depart to bear and rear their offspring with no further parental investment from the lek participants (Plates 6.8 and 6.9). Leks are well-developed among several species of North American grouse (Wiley, 1973), many African antelopes (Buechner and Roth, 1974; Jarman, 1974), and at least one species of bat (Bradbury, 1975) and fruit fly (Spieth, 1968).

Characteristically, a very small number of males is responsible for the vast majority of matings in such species. As expected, the resulting intense sexual selection has produced extremes of male ornamentation and gaudy display. In most species, females mate preferentially with males occupying particular, usually central, locations within the lekking gounds. Females are apparently attracted to males occupying these favored sites; however, unlike the polygyny threshold model described above, they do not derive any direct benefit from the real estate itself, since no feeding or nesting occurs on the lek. But because males compete actively for these preferred mating territories, females that choose the winners are likely to receive "good" genes in return; this is an adaptive mate selection strategy, given the mating system of the species.

The situation is a bit more complex from the males' viewpoint: Individual males who are successful in lek competition clearly profit from the system. On the other hand, unsuccessful males might seem better off, more fit, if they displayed solitarily, thus avoiding competition from neighboring males. Promiscuity in general and communal displaying in particular are most likely to evolve when one sex makes a minimum parental investment, often nothing beyond its gametes, not even defense of a territory or protection against predators. In such cases, the sex investing more, usually the female, would be selected for choosing the best of all possible gametes, and the

PLATE 6.8 MALE SAGE GROUSE DISPLAYING AT THEIR LEKS, WHILE SEVERAL FEMALES OBSERVE THE ACTION. (PHOTO BY H. WILEY)

PLATE 6.9 CLOSE-UP OF DISPLAYING MALE SAGE GROUSE. The two large egg-like structures are patches of bare skin that inflate with air, contributing to a loud booming sound made during courtship. Such extreme development of courtship is a product of sexual selection. (PHOTO BY H. WILEY)

females could best accomplish this by encountering all the males together so as to choose most efficiently among the rivals. Thus, female insistence upon convenient comparison shopping would maximize female fitness and this could in turn select for communal displaying by their potential mates. In addition, communal displays are often quite spectacular and can be seen and heard from a great distance, presumably increasing the advertising effect above that which a solitary male could generate. The disadvantages of competition with others may also be lessened if the successful males are related to those that are outcompeted. Thus, by contributing to the attractiveness of the lekking grounds, even unsuccessful participants may be enhancing their inclusive fitness, i.e., behaving in accord with an adaptive strategy (see Chapter 4). Finally, no individual is doomed to a continuing subordinate position within a lek: Dominant animals are eventually replaced from the ranks of the younger and less successful apprentices. Therefore, the system often has its own rewards for those that operate within it.

Alternatively, in some situations individuals may adopt particular strategies that enable them to beat the system. Male prairie chickens, a species of grouse, typically expose their bright air sacs and begin loud booming vocalizations within a few minutes of their arrival at a lekking ground (Plate 6.10). However some individuals remain inconspicuous, thus resembling females, and are courted briefly by the resident males. Only after reaching a desired location within the lekking ground do they show their true colors and begin typical male displaying. A similar strategy is adopted by certain male elephant seals, a harem-forming species: Sexually selected males are highly aggressive and have a large elephantine proboscis (see Plate 6.1). In contrast, some young males have small noses. They remain inconspicuous and avoid threatening the other males. This low profile strategy enables these young males to sneak into a harem disguised as females and occasionally achieve successful matings (Le Boeuf, 1974). Other individual strategies include intercepting females as they depart the harem (most females are already inseminated by this time, but some may not be); challenging a male harem master only after he is already

PLATE 6.10 MALE PRAIRIE CHICKEN IN COURTSHIP DISPLAY. Males of this species (closely-related to the sage grouse, Plates 6.8 and 6.9) occasionally obscure their sexual identity so as to gain admittance to a lek. (PHOTO BY H. WILEY)

exhausted from a long fight; and moving from one harem to another, remaining in the one that offers least competition from other males.

In one species of fish, the Gila topminnow, two distinct kinds of males can be identified. Large, dark males aggressively defend territories and engage in extensive courtship with females. Tiny, light males do not defend territories, are not aggressive, and engage in significantly less courtship; they sneak quick copulations from females (Costantz, 1975). Significantly, these small males possess a relatively longer intromittent organ. Within this fish family, body growth ceases when sexual maturity is attained, so adult male topminnows adhere to one or the other of these two discontinuous types, an example of a physical and behavioral polymorphism. As with most strategic decisions mediated by natural selection, the significant factors determining adoption of one strategy or the other can be revealed by a cost-benefit analysis. The large, aggressive strategy results in greater frequency of female encounters, an undisturbed mating environment, and increased competitive ability. On the other hand, the small, sneaky males become sexually mature at an earlier age, waste less time in aggressive encounters and may be less susceptible to predation. Ecological factors confer adaptive advantage to one strategy or the other depending upon local conditions; unpredictable environments may select for the quick-maturing, sneaky strategy, while constancy may favor the alternative. Similarly with high versus low predation pressure. Alternation of these conditions can maintain individuals of both strategies in the same place at the same time.

It is also true that the mating system in which members of one sex are most fit may well differ from that which maximizes the fitness of the other. In fact, such conflicts in optimum strategy should occur often, although they have not yet been studied in detail. One exception concerns the polygynous mating system of yellow-bellied marmots (Chapter 3) in which males are most fit when mated to a large number of females, whereas females are most fit when mated monogamously (Downhower and Armitage, 1971). In this case, the outcome appears to represent a compromise of optimum male and female strategies, with most social units being intermediate in size. Theory would predict aggression among the females in such a mating system, if an aggressive female can thereby evict other females, thus enhancing her fitness. This prediction remains to be evaluated.

In fact, other predictions can also be made; all are real predictions in that I am not familiar with any evidence bearing upon their validity. If the male's fitness is enhanced by the maintenance of a large harem, he may be selected to discriminate against an aggressive female, since in pursuing her own fitness by excluding other females, she would ultimately reduce his. If the male controls a resource of sufficient reproductive value to the female (his sperm or his habitat), the female may then be selected to reduce her aggressiveness if this increased her acceptance by the desirable male. By so doing, she would ultimately increase her fitness over what it would be if she were monogamously mated, but to a male offering fewer additional reproductive advantages. So, the relative quality of reproductive resource commanded by one mate should influence the strategic leverage effectively exerted against the other.*

Finally, under such circumstances, the mate favoring monogamy might be selected for a type of deceit, intrasexual amicability when the polygamously inclined mate is watching and aggressiveness when the mate is not. A similar argument could be made for other aspects of social interactions; e.g., mates of a monogamous species would be selected for minimal interest in copulations outside of the pair-bond, if such activities increase the likelihood of desertion by the mate. On the other hand, excessive readiness to desert as a result of a mate's peccadillos would itself be selected against, both because the roving mate would likely be more fit if associated with a more forgiving partner and because premature desertion would lower the probability of offspring success. And so it goes.

A mate seeking several reproductive partners would be most fit if these partners were minimally agressive, maximally altruistic, toward each other. This could be achieved by choosing mates who have a genetic interest in each other's reproductive success; ideally, siblings. This prediction also remains largely unevaluated. However, sociobiology relies heavily upon behavioral consequences of genetic relatedness, and we can expect such concerns to repay future research.

It is of substantial interest, for example, that in the two documented cases of simultaneous polyandry among animals, turkeys (Watts and Stokes, 1971) and Tasmanian native hens (Maynard Smith and Ridpath, 1972), the cooperating males are brothers. In such a system, the males lose some fitness since each may be excluded reproductively by the other. On the other hand, they gain something even in that eventuality, since by sharing one-half of their genes with their brother, they share

---

*No pity should be wasted on the manipulated mate, however, since given the natural circumstances, the acquiescing partner is in fact increasing its Darwinian fitness over what it would be if it failed to do so. It is employing an adaptive strategy appropriate to its particular situation.

171

one-fourth of their genes with his offspring, their nephews and nieces. At best, a given male may be a father; at worst, an uncle. The average return from such a strategy apparently exceeds the return provided by a monogamous mating in which, although paternity may be assured, fewer offspring are produced; trios are more successful reproductively than duos. However, in the absence of genetic relatedness among the males, this system would not produce an adaptive strategy.

A traditional view of reproductive behavior is that success is in the interests of all parties, thus making the production of offspring a heart-warming, cooperative affair. In contrast, the evolutionary perspective suggests the potential for conflict, because it emphasizes fitness of individuals rather than groups, species or even pairs. A world in which virtually every individual is genetically distinct is one in which substantial disagreements between indivduals would be expected, with each selected for maximization of its *own* inclusive fitness. Therefore real possibilities exist for conflicts of interest between individuals of the same sex (males versus males and females versus females), between mates (male versus female), and even between parents and offspring. The preceding pages have documented some of these conflicts and suggested some of the possible strategies, particularly with regard to the initiation of reproductive behavior, the prelude to breeding itself. The next chapter will continue this analysis into the later stages, those involving the actual production and care of offspring.

S E V E N
Strategies of Parenting

Animals do not seem to have much choice in the realm of parental behavior. Elephants can produce only one offspring at a time and they must provide a great deal of care and attention to that calf. Similarly, meadow mice have no alternative to the frequent production of large litters with minimal investment in any one offspring. We can consider such characteristics of each species as God given or somehow otherwise imposed upon living things, we can see them as random aggregates, the results of some cosmic dice game, or we can interpret living things as the products of natural selection. This latter perspective permits further insights and predictions. Each species is seen as having arrived at its particular constellation of reproductive behaviors as a result of the continual evolutionary testing of alternative strategies. Of course, the physical nature of each species, for example, its size, shape, kind of teeth, and pattern of hand or paw, cannot be separated from its behavioral repertoire: these are intimately and uniquely connected for each species. Evolution assesses the total phenotypic package presented by each individual of each species. Because a species is composed of a large number of individuals, each can be considered to have proposed a large number of alternative strategies and combinations of strategies, some of which were more successful, more fit, than others. In this sense, they have "choices".

Elephants and mice are different species, each with its own

set of parental strategies. These strategies make a coherent, well adapted bundle in each case; neither one is better than the other. Indeed, all successful species represent bundles of adaptations whose phenotypes and underlying genotypes are good enough to have kept them in the running. This chapter will consider parental strategies in terms of the major points at which choices are made by each species in the evolutionary race.

How to Partition Investment in Males versus Females

OK, so you're going to breed and you've already decided how, when, where, with whom, and in what social system. Now, do you want boys or girls? This may sound like an absurd question when stated in terms of an individual decision, but in fact the sex ratio of a species is a perfectly good biological characteristic, and there is every reason to expect it to be subject to natural selection. And of course, any characteristic of a species is simply the sum of the characteristics of all its component individuals, with evolution operating upon the choices made by each individual.

Most animal populations have a sex ratio that is very close to 1:1 or equal numbers of males and females. This is attributable to the genetics of sex determination: in mammals, for example, males produce two kinds of sperm, X and Y. If an X sperm fertilizes the female's egg, a female offspring is produced; if a Y sperm, a male. As a *proximate* explanation for equal sex ratio, this is fine. But consider that the above system will produce equal numbers of each sex only if, for example, males produce equal numbers of X and Y sperm and if the two are equally likely to achieve fertilization. In fact, there are numerous ways in which the sex ratio might be varied. Rather than considering the sex ratio as an immutable property of living things, it is more profitable to attribute it to the *ultimate*, adaptive significance of producing males and females in equal numbers, in other words, to the consequences in terms of inclusive fitness of this particular decision by the individuals making up the species in question.

Persistence of an equal sex ratio within any species implies a self-correcting system that prevents deviations in either direction (Fisher, 1930). Thus, imagine a monogamous species in which a population has developed an excess of males. Breeding individuals now face the choice of adding males or females to this system. Given that the existing unbalanced sex ratio has resulted in some nonbreeding males, the production of additional males, some of which would probably be reproductively excluded, would be a less fit strategy than would production of females, all of whom would likely breed. Optimal parental strategy would therefore be to produce females under these conditions. As the number of females begins to exceed the males, optimal strategy switches to the production of more males. The resulting system oscillates around equal numbers of each sex.

The situation is not materially changed if we introduce polygamous mating. Imagine a polygynous species in which one male mates with ten females. Nine males are therefore excluded for every one that breeds. If each female is worth one unit of fitness to her parents because each female breeds, then each of the nine excluded males is worth zero. However, the harem master is worth ten; his reproductive contribution to the next generation is exactly equivalent to the sum of his females. On balance then, the production of males versus females still generates equivalent parental fitness, although female offspring represent a more conservative strategy (all females breed) whereas the production of males is more risky; only one in ten will breed, but that one generates a payoff that exactly counterbalances the other nine males who do not succeed. Once again, any deviations in the sex ratio automatically set up counter pressures for production of the under-represented sex. Actually, it is not so much the *numbers* of offspring of each sex that will be equalized by natural selection but rather the parental investment in each sex (MacArthur, 1965; Hamilton, 1967; Leigh, 1970). For example, if sons were twice as expen-

175

sive to rear as daughters, parents would be selected for producing half as many sons.

This argument concerning optimal sex ratio in polygamous species applies only when parents have no information concerning the future competitive abilities of their offspring. However, there is some reason to believe that parents are not as ignorant as all that. Among many social species, primates in particular, the offspring of dominant females are likely to be dominant themselves, in part because they receive support from their dominant mother during aggressive encounters while they are growing up and perhaps in part because of genetic factors conferring greater likelihood of social success (Plate 7.1). In addition, healthy females of any species are more likely to produce healthy offspring that should in turn be more likely to succeed in social competition with their peers. In other words, a male offspring produced by a superior female would have a greater than average probability of being the one in ten that is successful. Given that the variance of male reproductive success is generally higher than that of females (see Chapter 6), the above considerations should select for differential strategies with regard to preferred sex of offspring: Produce males when you are dominant and/or in good physical condition; produce females when you are subordinate and/or in relatively poor condition (Trivers and Willard, 1973). Selection favors the risky strategy if the dice are loaded somewhat in your favor; otherwise, it is best to play it safe.

This is a striking prediction. What is more striking is that it is supported by the available data for most vertebrates, including pigs, sheep, mink, seals, deer, and humans. Stress during pregnancy, for example, induces higher mortality among male than female fetuses, in both deer and humans. In the past, this has been attributed to some generalized weakness of males; sociobiology now suggests a possible ultimate reason for this proximate factor. We can also predict that the adaptive significance of varying the sex of offspring will increase with increase in variance of reproductive success within either sex; i.e., highly polygynous seals should be more concerned with cor-

PLATE 7.1 FIFI, A HIGH-RANKING ADULT FEMALE CHIMPANZEE AT THE GOMBE
STREAM RESERVE, HOLDS HER 2½ YEAR OLD SON. The offspring of dominant
mothers tend to be dominant themselves when they become adult. (PHOTO BY
H. KLEIN)

177

relating sex of offspring with maternal condition than should monogamous songbirds.

There are numerous proximate mechanisms available with which parents could influence the sex of their offspring. As already mentioned, males could vary the proportion of X and Y sperm produced, and/or their viability, as a function of the female with whom they copulate. Females have more options, and appropriately so, considering their greater investment in the consequences of fertilization. Females could alter the chemical environment of the vaginal tract to favor the appropriate type of sperm; among humans, for example, male-causing sperm (Y) are relatively more viable in an alkaline medium, and female-causing sperm (X) are more viable in an acidic medium. The option of influencing the sex of one's offspring is being exercised increasingly by humans. There is no reason why animals couldn't do this naturally. Again, no volition is required; a correlation must only exist between physical condition and chemical secretion within the vagina. If that correlation is adaptive, it could evolve. Beyond this, females could alter the permeability of their eggs to X- versus Y-bearing sperm and/or induce early mortality in a fetus of inappropriate sex. Maternally induced abortion would be a relatively inefficient technique but, if used at all, it would minimize wasted investment if it occurred early in pregnancy. Significantly, this is when spontaneous abortions are more common, both in human and nonhuman animals.

Among a few species, the sex of an individual is *not* fixed at birth but can be changed at will; this peculiarity is also in accord with fitness considerations. Such biological transvestism has been well studied in a Pacific reef-dwelling labrid fish (Robertson, 1972). This species forms harems composed of a single male and many females. The females are kept female by being dominated by the male. When the male is removed, the top-ranking female quickly changes sex and keeps the others female by aggressively dominating them. Clearly, a male harem *master* is more fit than a female harem *member* because he mates with each female in his harem whereas each female mates only once. This explains why females compete aggressively to

178

become *the* male. However, even a female harem member is more fit than she would be as a subordinate male, who does not mate at all. This explains why subordinate females remain female.

How Much Should Be Invested in Each Reproductive Effort?

Evolutionary theory suggests that living things should devote themselves totally to the production of offspring. And in a real sense they do. Thus, such maintenance activities as eating, defecating, resting, scratching, preening, or whatever are ultimately of importance only as they translate to inclusive fitness. The same applies even more directly to social activities such as fighting, courting, defending territories, and avoiding predators. In fact, true hedonism, like true altruism, probably has no place in the biological world. Insofar as animals do something because they enjoy it, that very enjoyment is interpretable as an evolutionary strategy inducing living things to function in a way that ultimately enhances their fitness. In other words, organisms are programmed to seek rest when they are weary or food when they are hungry or to scratch when they itch, because to some extent these satisfactions are adaptive. Animals are selected to seek situations that enhance their fitness. (For an admittedly speculative extension to human behavior, see the discussion of "Why is Sugar Sweet?" in Chapter 10).

Thus, the answer to the question, "How much should be invested in reproduction?" is simply, "As much as possible, consistent with the maximization of inclusive fitness." Although this answer provides considerable analytic and predictive insight, it is also deceptively simple. There is every reason to believe that animals are selected for that level of reproductive effort that maximizes individual inclusive fitness, *during their entire lifetime of potential effort.* However, much variation remains possible with regard to how reproductive effort may be partitioned during that lifetime. The factors influencing this strategic decision are of major interest to evolutionary ecologists (e.g., Emlen,

179

1972; Pianka, 1974; Ricklefs, 1973), and the possibility of incorporating these ideas into social science raises exciting prospects for a true biological theory of parental behavior and the family.

Although animals have a diversity of parental investment strategies available to them, the options are not unlimited. Thus, if one elects to produce a very large number of offspring, there are usually restrictions in terms of the amount of investment in any one. The situation resembles what mathematicians describe as a zero sum game; that is, divide it up as one wishes, there is only a certain amount of "stuff" to go around. Parents can put their egg-producing stuff into many eggs, but in that case each egg will have to be small. Similarly, if many offspring are produced, then each one will also have to get relatively little parental attention, because there is only a certain total amount of time that parents can spend on their offspring. On the other hand, if parents are willing to produce a smaller number of offspring, then they can make each one larger and/or spend more time in parental care and attention. An alternative but equally accurate way of phrasing this would be: If parents insist upon producing offspring each of which is large relative to the parents and/or which requires a large investment of time and effort, then the parents have no choice but to produce fewer of them.

Which basic strategy will animals choose? Clearly, different species make different decisions, depending upon their genetic makeup and the ecological situation they face. For example, species such as most invertebrates and cold-blooded animals that rely heavily upon genetic coding of behavior require little if any parental instruction and will reproduce in large batches. In contrast, species whose young require substantial care and/ or training will tend toward smaller broods or litters with relatively prolonged juvenile dependency, as in most birds and mammals. Of course, within these larger categories much variation is to be expected, which is itself adaptive. Thus, birds that are helpless at birth and require feeding by the parents, tend to produce smaller clutches than do species whose precocious young are capable of obtaining their own food immediately

180

after hatching. In contrast, primates, including humans, have a prolonged period of infant dependency and, accordingly, singleton births are the rule.

Predation also experts a powerful influence. If the loss to predators is high and there is little that parents can do to increase the survival opportunities of any one of their offspring, then their best strategy is to generate lots of "cannon fodder", for example, the millions of eggs laid by a spawning codfish. But if offspring can be made relatively immune to predators, then another option is open; for example, a wildebeest calf can outrun certain predators when it is only a few hours old. Of course, in order to produce an offspring who is so precocious, mother wildebeest must expend considerable metabolic energy in the fetus, so only one is produced at a time. Adaptive correlations of this sort even exist within a single species. Certain European lizards produce large clutches of small eggs on the continent and small clutches of large eggs on Mediterranean islands (Kramer, 1946). The mainland population faces heavy predation, so parents are most fit if they generate enough offspring so that a least a few can survive. Predators are insignificant on the islands, but water is scarce, so parents under these conditions are most fit if they provide each young lizard with a sufficiently good start, i.e., large size, although they must sacrifice sheer numbers in doing so.

These two basic strategies are different extremes of a continuum, rather than discrete, either-or dichotomies. On the other hand, they do exemplify different fundamental patterns of great significance for sociobiology in general and parental behavior in particular. At one extreme are those species whose reproductive strategies emphasize the production of large numbers of offspring. These are the *r-selectionists*, so named because they maximize r, the population ecologists' symbol for the rate of population increase. This must be distinguished from the other r, the coefficient of relationship employed by population geneticists and discussed in Chapter 4. The rate of population increase is simply births minus deaths and r-selectionists have high birth and death rates.

Alternatively, a population can still be equally fit if it has low birth rates, as long as death rates are comparably low. This is the strategy exemplified by wildebeest, by primates and by island-dwelling lizards. In such cases, population size generally tends to be close to the number that the environment can support, its *carrying capacity*. Selection therefore acts primarily to increase the competitive abilities of the smaller numbers of offspring that are produced; this is functionally equivalent to reducing death rates. This emphasis upon quality rather than quantity tends to enhance the precision of ecological adjustment of the species within its niche, thus increasing the carrying capacity, symbolized by K. Species that elect this type of strategy are therefore designated *K-selectionists*. It is important to emphasize that r-selected and K-selected species are both functioning to maximize fitness of their constituent individuals. They are simply employing different proximate routes to the same ultimate end, because all sexually reproducing species end up with an average of two successful offspring for each two parents.

Table 7.1 summarizes the major characteristics of r- and K-selected species. These are not absolute criteria. Thus, mice are K-selected relative to oysters but r-selected relative to elephants. In general, however, large animals such as humans, eagles, elephants, and whales tend to be K-selected, and small animals tend to be r-selected. Parental care is better developed in K-selected species, consistent with the emphasis upon quality over quantity. One thesis of the present book is that social behavior constitutes a primary set of biological enabling devices, directed ultimately toward maximization of inclusive fitness for each participant. Most of the various advantages previously adduced for animal social systems relate generally to enhancement of fine tuning associated with adaptation. Accordingly, it is consistent that the social systems of K-selectionists tend to be relatively well developed, often organized around complex kinship networks, as compared to r-selectionists that tend toward either asocial living or loosely aggregated swarms, schools, or herds.

K-strategists tend to have longer lifespans and to delay maturity, whereas r-strategists live fast, love hard, and die

TABLE 7.1

SOME ASPECTS OF r-SELECTION AND K-SELECTION[a]

		r-SELECTION	K-SELECTION
Characteristic of the environment	Climate	Variable	Constant
	Availability of resources	Unpredictable	Predictable
	Habitat	Transient	Stable
Characteristic of the population	Mortality	High (independent of population size)	Low (often buffered by population, i.e., high when population is high, low when population is low)
	Competition (both within a species and between species)	Weak (unimportant)	Strong, (important)
	Energy utilization	Emphasizes quantity	Emphasizes quality
	Population size	Varies dramatically	Steady (close to carrying capacity)
Characteristic of the individual	Life span	Short	Long
	Body size	Small	Large
	Age at reproduction	Younger	Older
	Number of offspring	Many	Few
Characteristic of the social system	Parental care	Poorly developed	Well developed
	Social grouping	Poorly integrated i.e., female with young, aggregations, and survival groups)	Well integrated (i.e., family groups, often extended)
	Altruism	Rare	Common

[a]Modified from Pianka, 1970 and Wilson, 1975.

young. If lifespan is long and reproductive success increases with experience or, at least, does not greatly decrease, then optimal strategy is to breed in successive years (*iteroparity*). By contrast, short-lived, *r*-selected species tend to be more fit if they put maximum effort into each bout of reproductive activity; because there is little future reproductive potential, there is little reason to hold back. The extreme of this strategy is to breed only once, *semelparity*, often in a suicidal burst that has indelicately been described as a Big Bang (Gadgil and Bossert, 1970). Ecological aspects of semelparity versus iteroparity have received several detailed mathematical treatments (Cole, 1954; Anderson and King, 1970; Gadgil and Bossert, 1970; Fagen, 1972).

As expected, dispersal is generally well developed among *r*-selected species, whose rapid reproduction tends toward local overpopulation. It is adaptive for these rapidly produced offspring to leave their area of birth, thus both reducing competition with their parents and siblings, and also enhancing their likelihood of discovering a vacant habitat. Once there, the *r*-strategists' capacity for explosive reproduction renders them perfect colonizers. For example, meadow mice, *Microtus pennsylvanicus*, are adapted to wet grasslands, such as are produced behind abandoned beaver dams. They breed rapidly and disperse readily, each individual thereby increasing the likelihood of discovering and populating other transient habitats. They are opportunists, *r*-selected relative to the deer mice, *Peromyscus maniculatus*, that are adapted to stable, deciduous forests and produce smaller litters that are less likely to disperse (Christian, 1970).

Humans, of course, are *K*-selected as animals go. However, one of our most serious problems is that we are more *r*-selected than is good for us these days. This is not surprising, because our biological natures were essentially established during the millenia preceding advanced culture and technology. Thus, during the 99% of human history when our sociobiology evolved, we probably were not particularly abundant. Vacant habitats were almost certainly available and, although parental care was doubtless important, mortality was also very high, thus

184

selecting for a high birth rate or at least a higher birth rate than is appropriate today, given that modern medicine has lowered the death rate so dramatically. It is interesting that as human societies approach or exceed carrying capacity, optimum reproductive strategies may be shifting to quality rather than quantity; i.e., producing a smaller number of offspring but providing them with increased advantages such as a college education and good medical care, all of which ideally enhance their competitive potential. The *demographic transition,* fewer children per family in industrialized societies, is a switch to K-strategy, whether it is potentiated by genetic or cultural factors or some ineffable combination of the two.

The concept of r- versus K-selection sheds considerable light upon the sociobiology of parental strategies, but it doesn't explain everything. Imagine that it is late afternoon and a mother robin must decide whether to make one more foraging trip for her young brood. What factors will she have been selected to take into account? Probably the following: How hungry are her offspring? How tired is she? What are her chances of getting another worm? What are her chances of meeting a cat? Is this her first brood? Her last? Almost certainly, many other considerations are involved, but in the general case they boil down to basic issues of fitness. The act of making one more foraging trip, or doing almost anything, carries with it a cost in terms of reduced likelihood of survival by the parent, symbolied by s. This reduced survival decreases parental fitness by a factor $s(p+f)$, where p is the reduced success of the mother robin's present brood if she is unable to continue investment in them, i.e., she dies in the foraging attempt, and f is the mother's future reproductive success, which is of course sacrificed if she sacrifices herself. At the same time, each additional bout of worm hunting enhances the mother's fitness via an increased likelihood of the success of her offspring. This, in turn, varies with the quality and quantity of food she will obtain, the number of offspring, the extent of their need, etc. Mother robins will be selected for one more foraging trip when conditions are such that the consequence of failure, $s(p+f)$, is less than the increased fitness she gains by success. In other words, the behavior should occur if benefit exceeds cost. For a more elegant treatment of this theme, see Williams (1966b).

Analysis of this sort could relate to any behavior, but it seems particularly appropriate with regard to parental strategies, where the alternatives are often especially clear-cut. It also suggests numerous predictions. For example, as f declines, parents should be selected for greater willingness to perform the analog of getting one more worm. In other words, the lower an animal's future reproductive potential, the more inclined it should be to invest in any offspring at hand.

Alpine accentors are small, ground-nesting birds that inhabit the grassy meadows above timberline in the Alps. Breeding takes place during the very short summer and it must be done quickly if it is to be done at all. The chances of successful re-nesting decline rapidly as the season progresses. Therefore, parents should be increasingly attached to their young as these offspring grow older, since their year's reproductive success would become increasingly tied up with the fate of these offspring. Many animals, including accentors, show a *distraction display* when predators threaten their nest; perhaps the best known in America is the broken wing display of the kildeer, in which the parent appears to feign injury, thus luring the predator away from its nest. This is done at some risk, cost, to the parent but presumably with compensating benefit in terms of enhancing the likelihood of survival by the helpless nestlings (Plate 7.2).

As predicted, alpine accentors engage in distraction displays of increasing intensity as their offspring grow older (Barash, 1976d). The older the offspring, the more they represent in terms of parental reproductive success and parents are therefore willing (selected) to undergo greater risks on their behalf. Of course, investment should eventually decline, in some cases rather abruptly, when the offspring become sufficiently competent to survive without such parental risk, i.e., the benefit declines and/or the parents become increasingly prepared to invest in new offspring, i.e., the potential cost of failure increases. A parental defense pattern of this sort has been described for several other species (Kruuk, 1964; Lemmetyinen, 1971; Sparr,

186

PLATE 7.2 INJURY-FEIGNING DISTRACTION DISPLAY BY TWO PARADISE CRANES. Such behavior tends to lure predators away from the nest site, but it also subjects the parents to increased risk of predation. (NEW YORK ZOOLOGICAL SOCIETY PHOTO)

1974) but without apparent recognition of its adaptive significance. And a similar pattern may also apply to human behavior. (Chapter 10).

Western Washington is the breeding ground for two subspecies of a common small bird, the white-crowned sparrows, *Zonotrichia leucophrys*. The two forms, *z.l. pugetensis* and *z. l. gambelli* look almost identical. They differ principally in that *pugetensis* is adapted to long growing seasons; it breeds from Seattle south to central California, invariably renests if its eggs are destroyed, and typically breeds two or three times per year. In contrast, *gambelli* is adapted to the short growing seasons of high latitude environments. It breeds from British Columbia into Alaska, reaching the state of Washington only in the high North Cascades. *Gambelli* accordingly breeds only once per sea-

187

son and does not renest if its clutch is destroyed. The season is too brief to permit a second try. *Gambelli* parents literally have all their eggs in one basket whereas *pugetensis* parents can lose one clutch of eggs and still have another one or two chances at reproduction. According to sociobiologic theory, *gambelli* should therefore be more willing to incur risk on behalf of their offspring than should *pugetensis*. The intensity of parental defense in these two sub-species is being tested by evaluating their response to a model of an owl that is brought (via a clothesline pulley) close to their nests. As predicted, *gambelli* parents are significantly more courageous than are *pugetensis* (Barash, in preparation). Rephrasing the old cliche, he who fights and runs away is selected to do so insofar as he will be able to breed another day. Once again, and at the risk of further upsetting those social scientists who are totally committed to the role of experience in fashioning human behavior, some extrapolations to *Homo sapiens* will be presented in Chapter 10.

Who Takes Care of the Kids?

It is almost invaribaly true that if parental care takes place, it is done by the female. Why? Of course, one set of answers deals with hormones. The female sex hormones estrogen, prolactin, and progesterone in particular have been variously implicated in parental behavior. However, on the level of ultimate causation, we can ask why the female hormones have this effect, insofar as they do, rather than the male hormone testosterone. There are probably two sociobiological explanations for this. The first relates to the asymmetry of parental investment between males and females, discussed earlier (Chapter 6; and Trivers, 1972). Females generally have a much greater investment in their offspring; a greater stake in the young they have produced. Of course, if we consider the total of all offspring produced by an individual during its lifetime, then males and females have an equal interest. However, because of the differences in reproductive biology of males and females, the offspring of a female are likely to be concentrated in a smaller number of individuals with which the female can be closely associated, as opposed to the male's offspring, who may be spread across several females and be in different stages of development. In some cases such as the strictly monogamous eagles, geese, and foxes, male and female reproductive performance is one and the same. And significantly in these cases male and female parental care is also distributed about equally.

In general, the greater the parental investment, the more likely a parent is to care for its offspring or, similarly, the less likely it is to desert. Common sense suggests that past investment obligates each individual to continued future investment. But this is not strictly true. By similar logic, further American military involvement in Vietnam was justified as being somehow required so that previous losses would not have occurred "in vain". Parental investment, as with lives lost in Southeast Asia, is best viewed as what economists call "sunk cost". Thus, decisions for the future should be based upon options in the *future*, not past investment. In the case of parental investment, the relevant issues for each parent are those involving the consequences of alternative action for the fitness of the individuals concerned, not the amount of parental investment already accrued. The crucial question thus becomes "What is the best course of action *now*?"; that is, "What is the effect on my inclusive fitness of behavior A as opposed to behavior B?" and *not* "How much have I already invested?" (Dawkins and Carlisle, 1976). Of course, past accumulated parental investment may be of great significance in that offspring that have already received a large parental investment are likely to require less in the future—i.e., they represent a better investment, *in the present*. In this sense, the sex with the larger parental investment is likely to be more solicitous toward offspring; that sex is usually female.

The second major reason why females are generally more predisposed than males to caring for offspring is that females can be absolutely certain that they are related to their offspring, i.e., that their offspring are in fact, their own, whereas males cannot. This biological asymmetry applies to all species that practice internal fertilization, i.e., species in which eggs are fer-

189

tilized within the body of the female. Significantly, species engaging in external fertilization, such as many fish and insects, are notable for the extent to which males participate in parental care (Plates 7.3 and 7.4). Male stickleback fish, for example, defend a tunnel of love to which they attract several different females, each of which deposits eggs and then leaves. The male

PLATE 7.3 FEMALE WATER BUG *(above)* DEPOSITING EGGS ON THE BACK OF A MALE. He will now carry and aerate the eggs until they hatch. (PHOTO BY R. SMITH)

fertilizes them and then cares for the eggs. He is very likely the father of each of the offspring, and it is in fact more appropriate that he rather than any of his female consorts care for the eggs, because each female is related to only some of the progeny within any one nest.

In general, parental solicitude toward young is correlated with the likelihood of genetic relatedness. Combined with the

PLATE 7.4 MALE WATER BUG CARRYING EGGS ON HIS BACK. One egg can be seen hatching at the lower left. (PHOTO BY R. SMITH)

greater female investment in offspring, this may, in fact, be part of the reason why among mammals females invariably do the nursing. Considering that a new mother has just undergone the stresses of pregnancy and childbirth, it might seem reasonable and even adaptive for the males to take over the nursing, at least while she recovers her strength. But such a strategy is not adaptive for the new father, perhaps in part because he has no way of knowing that he is in fact the father. And given that the female is strongly selected to perform the appropriate parental activities because of her large investment, males are most fit if they parasitize that greater investment and genetic assurance of their mates: Males can get away with less parental investment, in part *because* females provide more.

However, don't be too readily seduced by the logical persuasiveness of evolutionary reasoning. Proximate and ultimate factors may often be intricately combined. In the case of male versus female lactation, for example, it is also quite possible that one factor predisposing females more than males is the simple fact that pregnancy and birth provide substantial opportunities for hormonal preparation of mammary tissue as well as precise timing for the onset of nursing itself. Indeed, the same hormone that produces uterine contractions is also responsible for the ejection of milk from the mother's breast.

Despite this caution against unitary explanations for complex behaviors, paternal behavior among nonhuman primates nonetheless reveals some interesting patterns that are consistent with the suggested importance of genetic relatedness (Plate 7.5). Siamangs are apes closely related to gibbons, and they inhabit tropical forests of Indonesia. The male siamang assumes a substantial parental role. In fact, he is the primary parenting adult during the day, carrying the juvenile and otherwise assisting it when necessary, then returning it to the female at night for nursing (Chivers, 1974). In contrast, male East African baboons are paternal only insofar as they tolerate infants (DeVore and Hall, 1965). Siamangs are strict monogamists; the male is almost certainly the father of the young siamang upon whom he lavishes so much attention. Baboons, on the other hand, live in troops within which male-infant relatedness is significantly

PLATE 7.5 MALE COTTON-TOPPED TAMARIN (A SOUTH AMERICAN MONKEY) WITH A BABY SCRAMBLING ONTO HIS BACK, WHILE THE FEMALE *(at right)* LOOKS ON. This species is monogamous, and the males can be considerably confident that they are the fathers of infants produced by their mates. Males accordingly show considerable parental care. (PHOTO BY J. SPURR)

more uncertain. It is interesting that among baboons, male solicitude for infants is most apparent when the males defend the whole troop against predators such as leopards or cheetahs; although the dominant male is necessarily uncertain about his paternity relative to any one infant, he can be reasonably confident of his average relatedness to infants in the troop taken as a whole.

But again, things are rarely as simple as we might wish them. Gibbons appear to be no less monogamous than siamangs, yet male gibbons appear to be significantly less paternal (Carpenter, 1940). Gibbon males are somewhat more involved in territorial confrontations with other males than are male siamangs. It is therefore possible that the paternal behavior of male siamangs derives from their having become liberated from extensive territorial patrolling and aggressive behavior. Cause and effect are one of the basic organizing principles of science and, indeed, everyday experience as well, but ecological and evolutionary thinking suggests increasingly that such discrete linearity may often be inappropriate to natural systems. For example, assuming that the above speculation regarding siamang versus gibbon paternal behavior is correct, why has this difference evolved? It could be that some characteristic of siamangs necessitates paternal child-rearing assistance and that, accordingly, they "make do" with minimal territorial hassling. On the other hand, perhaps some characteristic of siamangs predisposes them to a low level of physical male-male interactive aggressiveness, such that increased paternal concern derives secondarily. Whichever is primary, such tendencies are mutually reinforcing once they are initiated. Increased efficiency of adaptation A facilitates adaptation B, whereas B has a similar effect on A, obscuring which came first.

A recent study of the "Ecology of Paternal Behavior in the Hoary Marmot" revealed interesting variations in male behavior toward their offspring, reflecting a pattern consistent with the maximization of individual fitness (Barash, 1975c). These animals were found in two distinct kinds of social systems: isolated family units consisting of one male, one female, and their offspring of the last two years; and populous colonies extending over many acres and consisting of many males, many females, and their numerous offspring. Adult males in the colonial situation behaved as typical rodent fathers; they interacted very little with their offspring. Their social interactions were devoted almost entirely to the other adults. In contrast, adult males inhabiting the isolated families devoted considerable attention to their infants, engaging in mutual grooming, play-fighting, etc.(Plate 7.6).

This pattern is highly adaptive. Seen from the viewpoint of an individual male seeking to maximize his fitness, what can he do? Like all male mammals, he cannot nurse his offspring. He lacks the opposable thumbs of the siamang with which to carry his children. Since marmots are vegetarians, he also cannot efficiently provide solid food. After all, grasses are low in food value and therefore, inefficient to transport and, besides, they are easily available to any hungry youngster. If marmots fed

PLATE 7.6 ADULT MALE HOARY MARMOT "GREETS" AN INFANT. Males at isolated colonies show more paternal behavior than do males that have the opportunity of interacting with other adults. (PHOTO BY J. TAULMAN)

upon substances of high caloric value such as meat, then perhaps males could advantageously help provision their offspring. Significantly, male assistance in provisioning of infants is almost uniquely characteristic of carnivores: wolves, foxes, hunting dogs and hyenas. So what is left? Not being able to do much via their offspring directly, males in large colonies of marmots maximize their fitness by interacting with the other adults. Specifically, they attempt to solicit copulations with neighboring females, while maintaining a watchful eye and aggressive demeanor toward nearby males who have similar designs upon their females.

In contrast, isolated family males are in a different situation. By virtue of their social isolation, they need not fear encroachment of competing males upon their sexual prerogatives. At the same time, because there are no neighboring females, the recompense for roguery is greatly diminished. So they concern themselves with their offspring, essentially because they have

195

nothing better to do. The young marmots presumably receive some benefit from paternal solicitude, although it is not obvious how they profit in terms of fitness from their father's attention. The young animals may gain enhanced social competence through the added social experience. In addition, the male's extra attentiveness may provide added watchfulness vis-à-vis predators; the infants are often preoccupied with mutual play near their burrow entrance and are frequently warned of predators by a nearby adult, who may even nudge them underground.

The basic pattern suggested here is one of facultative paternal behavior (capable of being adaptively varied) constrasting with obligatory maternal responsibilities. This seems to be a general pattern among mammals, consistent with the expectations of sociobiology described above. For another example, the basic family unit among orang-utans in Sumatra consists of a mother, father and their offspring. In Borneo, males defend territories within which females and their infants are the central unit (Mackinnon, 1974), but the males are not incorporated into a close family unit. Predators and interspecific competitors are more abundant in Sumatra, so the most adaptive male strategy is to be a bodyguard for his family. Like the colonially living male hoary marmots, male orangutans in Borneo are more fit when they interact socially with other adults. This adaptive flexibility in male parental behavior is not restricted to mammals, although it is particularly well developed within this group. Among birds, for example, male long-billed marsh wrens mate polygynously. Early in the breeding season, they do not help feed their offspring; they are busy soliciting additional mates. However, when it is too late to start a brood and all available females are mated anyhow, they switch to a strategy of doting fatherhood and assist in provisioning their offspring (Verner, 1964).

Despite the temptation to place value judgments on these differences between male and female parental strategies, it should be pointed out that the essential selfishness of male parental behavior in these cases is in no way reprehensible. In fact, it is no more selfish than is female parenting; both repre-

196

sent strategies that maximize individual fitness, given the biology of maleness and femaleness, and the ecological constraints operating in each situation. Furthermore, polyandrous social systems in which males assume a substantial parenting role have been identified in a variety of species, e.g., seahorses (Fiedler, 1954), dendrobatid frogs (Sexton, 1960), jacanas, neotropical wading birds (Jenni, 1974), and phaparopes, small shorebirds (Hilden and Vuolanot, 1972). Significantly, these reversals of typical male-female parental roles correlate with reversals of typical male-female parental investment (Chapter 6; Trivers, 1972).

Songbirds are notable for the degree to which both males and females show parental care. This mutuality of behavior is correlated with high male confidence of paternity and also with the adaptive value of having two committed adults assisting in the rearing of young. For example, the high metabolic rate of young songbirds requires a large daily input of food, and their helpless nature means that they must be fed, because they cannot forage for themselves. The answer, of course, is monogamy, with both parents providing food. Among Antarctic penguins, breeding occurs on land whereas food can be obtained only at sea. Because the severe environment requires that one parent remain constantly on duty to prevent the egg from freezing, these birds mate monogamously and alternate in parental duties; a widowed penguin cannot rear offspring successfully. In short, animals share parenting when such a pattern is adaptive, but it must be adaptive for *each* participant.

Some species are more flexible in their parenting requirements. For example, bobolinks, common birds of American midwestern prairies, form secondary as well as primary mateships. A male helps his primary female to provision their offspring, but normally he does not aid his secondary mate. In turn, secondary females have evolved a variety of strategies that compensate for the absent father: They lay smaller clutches, thus dividing the limited food among fewer mouths; they are less fussy about potential food items and also capitalize upon insects found closer to the nest; and they are more likely to cease feeding undersized young in times of scarcity. In addition, when secondary females occasionally produce clutches that are exceptionally large, the male sometimes helps provision these offspring (Martin, 1974). Such flexibility in male parenting is also clearly adaptive, although it is also interesting to speculate upon the possible strategic ploy of secondary females in producing an oversized family in order to lure the male's attention; i.e., in their case, the best way to a man's heart may not be through his stomach but, rather, through an appeal to his fitness.

Care of infants need not be restricted to parents. In some cases, nonparental adults may assist the parents. We have already considered "helpers at the nest" in the discussion of altruism (Chapter 4). Such nonparental helpers tend to be closely related to the individuals being helped, suggesting kin selection. They also tend to be young, suggesting selfishness, in the sense of gaining experience in child care while they await the reproductive opportunities that will come with increased age. Nonparental care of young is particularly well developed among the primates, and the term *aunting* behavior has become part of the literature of primatology (Rowell et al., 1964; Plate 7.7). However, this implies both relatedness and female helpers, neither of which is necessarily true, so the term *alloparent* may well be preferable (Wilson, 1975).

Primate alloparents commonly groom infants, carry them extensively and may protect them from danger (Lancaster, 1971). An interesting distinction can be made between male and female alloparents: Females seem to direct their attention particularly toward the well-being of the infant in question, thus suggesting the possible learning and early expression of maternal skills. In contrast, male alloparenting tends to be more overtly selfish. Two particular patterns can be identified. In one, exemplified most clearly by hamadryas baboons, each male adopts, one might almost say kidnaps, juvenile females who later become part of his harem (Kummer, 1968). Among nearly all primates, adult females with infants are accorded a degree of social tolerance by the adult males. In the second form of allopaternal behavior, subordinate males take advantage of this inhibited adult aggressiveness by snatching up infants and in-

PLATE 7.7 JUVENILE LOWLAND GORILLA RIDING ON AN ADULT FEMALE (AN ALLOPARENT); ITS MOTHER IS OCCUPIED WITH A NEW INFANT. Adult male at right. (SAN DIEGO ZOO PHOTO)

terposing them between themselves and a more dominant animal (Plate 7.8). Such *agonistic buffering* has been described in detail among the Barbary Apes, which are really a species of macaque (Deag and Crook, 1971). It apparently is the nonhuman primate equivalent of "don't hit a man with glasses!"

For a concluding example relating alloparental care to other issues current in sociobiology, consider a schema for "Alternate Routes to Sociality in Jays" (Brown, 1974). The different species in this family employ social systems ranging from classic, all-purpose territories (scrub jay) to densely aggregated nesting colonies (pinon jay) on the one hand and communal breeding groups with substantial alloparental care (Mexican jay) on the other. Mexican jays are extraordinary in the extent to which only a small number of adults breed, assisted by numerous helpers who are often themselves adult. This species shows a behavioral syndrome typical of K-selection: low birth and death

199

PLATE 7.8 HAMADRYAS BABOON MALE CLUTCHING AN INFANT FOR SECURITY
WHILE ENGAGING IN DEFENSIVE BEHAVIOR TOWARD ANOTHER ADULT MALE
WHO HAS THREATENED HIM. (SAN DIEGO ZOO PHOTO)

rates, reduced dispersal, delayed maturation, and occupation of ecologically mature, stable habitats with population levels that approach carrying capacity.

Under such a regime, selection would likely favor the retention of young as helpers for the parents. This would be favored because of the fitness advantages that would accrue to both the parents and the helpers themselves. Retention of young also facilitates success in territorial defense, itself more important in K-selected populations, because larger groups are more successful in such defense. As selection favors the defense of high quality territories by kin groups, kin selection and finally even group selection could therefore begin to operate. Thus, parental strategies and K-strategies, operating in part through kin selection and possibly group selection, may ultimately create the social system characteristic of the species.

Should Parents and Offspring Disagree?

The drive to reproduce is evolution's handiwork and one of its strongest imperatives. Success of a parent is measured largely in terms of the reproductive success of its offspring and, because the offspring themselves also maximize their own fitness by reproducing successfully, it would seem that the interests of parents and offspring should coincide. Both should be selected for behaviors that promote success of offspring; i.e., success of the parents and success of the offspring appear to be one and the same or, at most, different sides of the same coin.

With this view, child-rearing is a process of mutual fitness seeking, with offspring and parent wonderfully united in a common goal, thus generating the heart-warming sagas of which television documentaries are made. Whatever conflicts may occur between parents and offspring are interpreted as either the unavoidable friction generated by an imperfect real world, or as a consequence of greater parental wisdom which ultimately rebounds to the benefit of the somewhat errant but simply misguided and headstrong youngster. For example, those parent-offspring struggles that are almost invariably associated with fledging in birds and weaning in mammals are commonly interpreted as mechanisms whereby the parent gently nudges the offspring toward independence and, thus, its own true self-interest.

There is considerable merit in this view, even beyond the emotional appeal of a world activated by a beneficent mother nature. Indeed, parental love is probably neither more nor less than an evolutionary strategem insuring that parental investment is sufficient to maximize parental fitness. Similarly, the love of offspring for parent is another word for an affiliation that maximizes offspring fitness. But as parent-offspring relations are rephrased in the evolutionary language of sociobiology, a potential for real conflict also becomes increasingly apparent. Thus, if parents and offspring are each ultimately seeking to maximize their own personal fitness, then they cannot be expected to agree entirely on an optimum strategy unless they are genetically identical. And in sexual species, parents and offspring are never genetically identical. In fact, there are numerous opportunities for parent-offspring conflict, the evolutionary bases for which will now be explored. See the pioneering work of Trivers (1974) for a detailed treatment.

Consider almost any species of mammal. An adult female is nursing her offspring, with both ultimately profiting by this exchange; the infant gains directly, the mother indirectly by enhancing the fitness of her offspring. At a certain point, however, the mother's best interests are served by investing in a second infant, with whom she will share the same proportion of genes (one-half) that she does with her present infant. In this sense, she has an equal interest in the success of each of her offspring; her fitness is served by the production of as many successful offspring as possible. The nursing infant, however, has a different view of things: It shares one-half of its genes with its mother and would likewise share one-half of its genes with a full sibling, but it "shares" all of its genes with itself. In other words, it is more interested in its own reproductive success than in a sibling's. Another way of viewing this is that it will share one-half of its genes with its own eventual offspring but only one-fourth with those of its sibling; therefore, each offspring is more interested in its own success than in that of its brother or sister. In contrast, the mother is equally interested in the success of all the offspring she can produce. Accordingly, insofar as an offspring's efforts to maximize its own personal fitness reduce the mother's ability to maximize her own fitness via the production and nurturing of another offspring, conflict between offspring and parent can be anticipated (Plate 7.9).

In our previous example, the mother was selected to cease nursing her current offspring at a certain point and to begin investing in the next. The nursing juvenile, however, would enhance its own fitness by continuing to nurse, rather than seeing that investment given to a sibling. Weaning conflict will therefore occur until such time as the interests of mother and infant again coincide; namely, when the cost to the mother of continued investment exceeds twice the benefit received by the feisty offspring at hand. Beyond this point, the infant's inclusive fitness is reduced by further decrements to the production of its own siblings, and both parents and offspring should agree that investment in the current offspring should be discontinued. Restating this: A parent should reject its offspring when the benefit of so doing for the parent exceeds the cost for the parent; offspring should accept this rejection only when the parental benefit exceeds twice the offspring's cost. Again, this is because $r = 1/2$ between parents and offspring; therefore offspring are only one-half as interested in parental cost as the parent is.

When the offspring is small, cost to the mother is small and for both individuals, benefit is great. Therefore, the mother wants to nurse and the infant wants to be nursed. However, as the infant gets larger, it becomes more expensive to maintain; it requires the investment of more units of maternal fitness. At the same time, it becomes increasingly feasible for the mother to invest in a new offspring, because the present one is increasingly able to survive without further investment by the mother. For its part, the offspring still wants the parental investment while the parent wishes to terminate. The offspring should acquiesce to termination of parental investment only when its own inclusive fitness requires it; when the cost in terms of the

PLATE 7.9 MONGOLIAN WILD HORSE FEMALE NURSING HER YOUNG. According to the theory of parent-offspring conflict, mothers and infants should agree on nursing, early in the life of the offspring. However, conflict should occur at a later stage, with the offspring seeking more investment (in this case, milk) than the parent is selected to give. This is supported by the observation that conflict over weaning is common in nature, e.g., as the colt matures, its efforts to nurse will be rebuffed increasingly by the mother. (NEW YORK ZOOLOGICAL SOCIETY PHOTO)

reduced success of the next sibling exceeds twice the benefit derived by the offspring itself from continuance of the investment. The expected correlation between age of offspring and fitness ratios is shown in FIG. 7.1, with the zone of conflict indicating the region of discordance between parental and offspring inclusive fitness.

The preceding discussion considered parent-offspring conflict over whether parental investment should be continued at all. The same analysis also predicts conflict over the amount of investment to be provided at any one time. Parents wish to invest in offspring so as to maximize parental return; they are

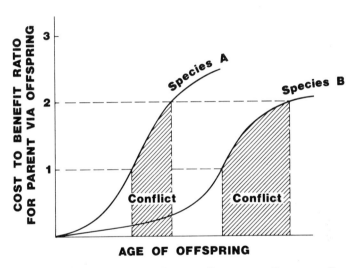

AGE OF OFFSPRING

FIGURE 7.1. PREDICTED PARENT-OFFSPRING CONFLICT REGARDING THE TERMINATION OF PARENTAL INVESTMENT. Parents are selected for terminating investment in their offspring when the costs of such investment are greater than the benefits, i.e., cost to benefit ratio exceeds one. In contrast, offspring are most fit if parental investment continues until the cost is greater than twice the benefit, i.e., cost to benefit ratio exceeds two. Therefore conflict should take place when the ratio of cost to benefit is greater than one but less than two. When less than one, parents and offspring agree to continue investment; when greater than two, they agree to terminate investment. Species A and B differ with regard to the rate of offspring development. Because A develops more rapidly, the zone of conflict is reached at an earlier age. (*Modified slightly from Trivers, 1974 and Wilson, 1975.*)

selected to maximize the difference between benefit and cost. Offspring, on the other hand, are only one-half as concerned about parental cost as are the parents themselves, because offspring are related to their parents as one part in two. Therefore, they seek a level of investment which maximizes the differences between benefit and *one-half* the cost (FIG. 7.2). Parents and offspring should thus disagree not only with regard to

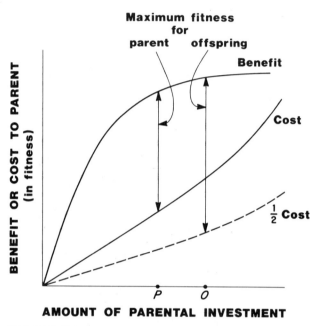

AMOUNT OF PARENTAL INVESTMENT

FIGURE 7.2. PREDICTED PARENT-OFFSPRING CONFLICT REGARDING THE AMOUNT OF PARENTAL INVESTMENT TO BE EXTENDED FOR ANY ONE OFFSPRING. With regard to any act of parental investment, parents are selected to maximize the difference between the benefits derived from that act and the costs imposed by it. This occurs when parental investment equals P. In contrast, offspring are selected to maximize the difference between the benefits derived from parental investment and one-half the costs, because they share only one-half of their genes with the parent. This occurs when parental investment equals O. Therefore, parents and offspring should be selected to disagree over the amount of parental investment to be provided, with parents preferring P and offspring preferring O. (*Modified slightly from Trivers, 1974.*)

206

when investment should be terminated but also about the amount of investment that should be provided at any one time, with offspring wanting to take more than parents are selected to give (Plate 7.10).

This approach illuminates a great deal of parent-offspring interaction. For example, because parents are equally related to each of their offspring, they would be selected for encouraging any altruistic behavior between two siblings such that the benefit derived by the recipient was greater than the cost suffered by the initiator. Similarly, they should discourage any behavior that harms one of their offspring more than it benefits the other. From the viewpoint of any one offspring, however, altruism would be advantageous only when the recipient sibling gains more than twice what the altruist loses. Thus, offspring should resist parental exhortations to assist their siblings. And

PLATE 7.10 ADULT FEMALE BURRO REJECTS HER FOAL, WHO IS ATTEMPTING TO NURSE. For these animals, living in Death Valley National Monument, rejection of this sort begins at birth, with the foal being progressively less successful in its nursing attempts as it grows older. Their arid environment probably increases the parental cost associated with nursing, thus displacing the curve of parent-offspring conflict toward younger offspring age. (PHOTO BY P. MOEHLMAN)

similarly, each offspring should tend to disadvantage a sibling as long as that cost is not more than twice the benefit derived by the initiator from its act. However, *parental* fitness is reduced by such behavior. Parents should therefore intervene to discourage excessive offspring egoism and selfishness relative to siblings, while also encouraging cooperation and sharing. Behaviors of this sort seem to occur among primates and some of the social carnivores.

However, our own species provides some of the most striking opportunities for verifying these and other predictions. For example, since parents share one-fourth of their genes with their nieces and nephews and their own offspring share only one-eighth of their genes with these same individuals, parents should encourage their offspring to play twice as nicely with their cousins as the offspring themselves would be inclined! The cogent theme of parent-offspring conflict in human behavior is further elaborated in Chapter 10. It is a novel concept and still perhaps a bit rough around the edges, but it is consistent with the way evolution should operate and with the way behavior actually does. Certainly, it provides a fresh, new perspective on the biology of sibling rivalry and the so-called generation gap.

E I G H T
Strategies of
Social Competition:
Aggression and Dominance

Cooperation and altruism are major themes in the evolution of social behavior. Nonetheless, the unpleasant fact remains that maximization of personal, inclusive fitness often involves asserting one's self at the *expense* of friends, neighbors, and even relatives. So long as individuals are genetically distinct, they can be expected to look out primarily for their own best interests. Of course, in some situations this may involve acts of extreme self-abnegation, as in sterile workers among the eusocial insects, discussed in Chapter 4. However, interactions among individuals are more often competitive, and rather feisty, especially among the vertebrates.

Competition occurs when two or more individuals seek access to a resource that is somehow important to the fitness of each and that is restricted in abundance such that optimal utilization of the resource by one individual requires that another settle for suboptimal utilization. In other words, if there is enough to go around, then there is no reason for competition—e.g., few animals ever compete for air. However, severe competition may erupt over food, water, nesting sites, and/or appropriate mates. We can identify two basic ways in which individuals compete for such resources, *scramble* and *contest competition.*

Scrambles occur when each participant attempts to accumulate and/or utilize as much of the critical resource as it can, without regard to any particular social interaction with its com-

209

petitors. If the resource is used up in the process, then the so-called winners of scramble competition are the individuals who have converted the largest part of that resource into copies of themselves, i.e., those that are most fit. Fitness in this case has been achieved by simply out-reproducing the competition, usually by being most efficient at garnering the resource in question. By definition, social interactions are excluded from this type of competition. If, on the other hand, competing individuals interact directly with each other and use the outcome of such interactions to determine access to resources, then contest competition is taking place. To the victor belongs the spoils, in contest competition. In scramble competition, the victor is simply the one that scrambles for the most spoils. This is an Easter egg hunt, in which every participant ignores every other and simply concentrates upon finding as many eggs as possible. In contrast, contest competition would be occurring if the participants first argued, fought, or somehow disputed among themselves, on the basis of which they decided who would look where, who would have first choice of the eggs collected, etc.

Aggression is the proximate mechanism of contest competition. It takes place when individuals interact with each other such that one of them is induced to surrender access to some resource important to its fitness. The exact forms of aggression vary widely, from intimidating displays and threats to actual fights. Just as animals ought to exert themselves to acquire important resources or enlarge their supply, thereby enhancing their fitness, they also ought to resist the loss of important resources, thereby avoiding decrements to their fitness. Accordingly, animals may respond to aggression by threatening back, fighting back, and, occasionally, signalling their submission and/or running away. All these encounter patterns are subsumed under the term *agonistic behavior*.

Social scientists have generally been inclined to view aggression, human aggression in particular, as inappropriate and undesirable behavior, essentially a form of neurosis (Montagu, 1968; Fromm, 1973). It is certainly true that aggression can be disadvantageously exaggerated, especially among humans in a

210

technological society with access to all the lethal weaponry our ingenuity has provided. Human aggression is even more maladaptive when it is performed by entire nations that have unimaginable destructive power at their disposal. Human manifestations of aggression may thus run largely counter to the maximization of fitness. The infliction of pain and suffering clearly deserve our opprobrium and indeed, defending aggression is like rooting for the guys with the black hats in a western movie. On the other hand, the evil here is probably not aggression per se but, rather, the uses to which aggression is put among human societies in which culture has imbued agonistic behaviors with the means of wreaking such havoc.

In fact, to some extent aggression appears to be a fundamental characteristic of nearly all living things, at least all that find it adaptive to engage in contest competition. The crux of a sociobiological analysis of aggression is to identify the role of aggression in the adaptive strategy of each living thing. Thus, we can analyze the form, context, and consequences of aggression in terms of strategic decisions, knowing that each decision has been and is being evaluated for the individuals comprising each species, in terms of fitness considerations for each alternative strategy.

Konrad Lorenz is the Nobel Prize-winning founder of ethology, the biological study of animal behavior. In his book *On Aggression*, Lorenz discussed various ways in which aggression is adaptive to those animals that engage in it; in fact, a literal translation of the original German title would have been "The So-Called Evil". Aggression is indeed widely distributed among animals and there is certainly abundant evidence that it has a substantial heritable component. It is also clear that aggression is influenced by a variety of experiential factors, especially early experience and the social context in which it is elicited. However this modifiability does not argue against the view that aggression tends to be adaptive when it does appear, in terms of both its specific form and its context. In adjusting behavior so as to maximize the fitness of each behaving animal, natural selection is often careful to ensure that the behavior occurs under

211

situations when it is maximally appropriate, when the correlations between experience and aggressive behavior are probably *themselves* adaptive.

The following chapter will attempt a sociobiologic approach to aggression, hopefully revealing some of the adaptive strategy reflected in these behaviors. One caution, however: If we choose to agree that aggression is adaptive this is not to say that we must consider it to be good. Aggression simply *is,* just like the big brain of a human being, the sticky tongue of an anteater or the shennanigans of a love-crazed prairie chicken. We are not required to like it or to dislike it, but we are well-advised to understand it.

When Should an Individual be Aggressive?

Evolutionary theory suggests a simple answer to the question, when should an individual be aggressive? The answer is consistent with the other strategic issues considered thus far: each individual should be aggressive when such behavior increases the individual's inclusive fitness. In other words, aggression should characterize situations in which contest competition is more efficient than scrambling. This is particularly true when resources are in short supply; for example, if there are ten hungry hyenas attempting to feed from a single zebra that offers enough meat for only five, then we would expect aggressive interactions among the competing animals. Alternatively, we might expect to see the *results* of previous aggressive competition, i.e., a hierarchy of social dominance among the competitors, on the basis of which the contested resource is apportioned.

Food is a common cause for aggression, largely because of its continual importance to animals. The physical organization of the resource is often very important. Aggression among otherwise peaceful baboons can be elicited by introducing a small amount of food, thus generating competition over the concentrated resource. Baboons normally eat grass shoots that are widely dispersed and therefore are best harvested by scrambling rather than contesting. Free-living Indian rhesus monkeys that obtain natural food from the forests are significantly less aggressive than are urban monkeys that inhabit temple grounds and compete for handouts (Singh, 1969; Southwick et al., 1976). The forest inhabitants are utilizing food that is widely distributed whereas the temple dwellers are seeking food that is available only in concentrated bundles. Clumping of a resource thus adds to aggression, both because it brings individuals close together, making competitive interactions more likely, and also because when it is clumped, a resource is more capable of being successfully defended, making competitive interactions more profitable.

Contest competition is therefore more likely to occur when a resource is clumped than when it is homogeneously distributed. Oceanic birds such as puffins, auks, murres, cormorants, and many gulls typically feed by plunge-diving into schools of herring and other fish. These resources are highly clumped but, because they also shift in location, it would not be an adaptive strategy to defend a patch of open ocean. In addition to the difficulty of establishing boundaries, there would be no value in competing aggressively for any one area unless the fish were there, and fish schools don't stay in one place for long. Accordingly, these birds do not defend foraging areas. They do, however, defend nesting sites that are also necessary for fitness and often limited in abundance (Plate 8.1).

As expected, species differ concerning which resources are worth competing over. East African baboons are particularly susceptible to predation at night. Accordingly, the area traversed by each baboon troop invariably includes one or several "sleeping trees", which provide essential nocturnal protection. In contrast, for the desert-dwelling hamadryas baboons, water is crucial and groups never travel very far from it (Altmann, 1974). Fur seals breed in large harems on rocky islands in the North Pacific. The males arrive on the islands first, whereupon they compete aggressively for status and a piece of real estate. The resource toward which competition is ultimately directed is the adult females that will comprise the harems of successful competitors. Since these females arrive

PLATE 8.1 THICK-BILLED MURRES ON THEIR CLIFF-EDGE NESTING SITES.
These animals do not defend their foraging areas on the open ocean, how-
ever, they do defend nest sites, which are in short supply (PHOTO BY R.
TENAZA)

from the ocean and are immediately herded by one proprietor
or another, waterfront property is an essential and highly valued commodity. Unsuccessful, subordinate bulls congregate in
upland areas with no beachfront and hence, no females.

Aggressive resource competition occurs when the resource is
limited in quantity and therefore worth fighting over (Plate
8.2). Establish a bird feeder during the winter, and it will probably attract many individuals, perhaps of several different
species, with considerable aggression occurring among them.
In contrast, the feeder will attract little attention in the spring
and summer, and those animals that do appear are unlikely to
fight. The simple reason is that food is abundantly available in
nature during these seasons; in addition, this food is often of
higher quality, e.g., insects, thus making it maladaptive to compete over a few paltry sunflower seeds. Aggressive competition
may therefore be strongly seasonal in occurrence, because the
resource itself may vary seasonally in terms of its density of dis-

PLATE 8.2 Two Hermit Crabs Fighting over a Snail Shell. These soft-bodied crabs rely upon empty shells to provide protection. As they outgrow their shells, they must move to progressively larger ones, thereby making appropriately sized shells a valuable resource, and one worth fighting over. (PHOTO BY B. NIST)

tribution and/or because it may vary in its desirability from one season to the next.

Mates are an important resource and, judging from the amount of aggression they generate, they are often worth competing for. In the temperate zones, seasons are clearly demarcated, resulting in distinct breeding seasons because animals are selected for reproducing at times that will maximize their fitness (Chapter 6). Intrasexual aggressive competition is therefore highly seasonal among most animals, because a mate is generally of value only when breeding is likely. The correlation of breeding and aggressiveness is particularly obvious among the ungulates whose mating season (rut) occurs in autumn; hikers are well-advised to give adult male bison a wide berth in late September and October, although bison are relatively placid

215

the rest of the year (Plate 8.3). The frequency of collisions between Alaskan bull moose and automobiles rises dramatically in
the autumn, and even freight trains are not immune!

For many species, increased crowding has important consequences for aggression. Increased crowding results in a decrease in the space available per animal and hence, an increase
in the frequency of encounters. This in itself could lead to an
increase in aggression by the simple mechanism described
above, relating aggression to clumping of resources. At the
same time, decreased space per animal reflects a more general
consequence of crowding: greater competition for the remaining resources. Crowding of captive rhesus monkeys results in
heightened aggression (Southwick, 1969), and this seems to be

PLATE 8.3 TWO ADULT MALE BISON *(one almost completely obscured by dust)*
FIGHT FOR DOMINANCE DURING THE RUTTING SEASON. Dominant animals
gain access to females, and hence, reproductive success. Accordingly, ability
to succeed in such encounters is strongly favored by natural selection, resulting in the large size and aggressive demeanor of bison bulls. (PHOTO BY D.
LOTT)

216

the most common vertebrate pattern. Invertebrates have not been carefully studied in this respect but, contrary to the usual vertebrate trend, male crickets are substantially more aggressive when kept in solitary confinement than when housed in groups.

Particularly because of the potential human implications, it is tempting to generalize about aggressiveness as a function of crowding, but the situation is far from clear. For example, increased social proximity among captive rats led to the emergence of a wide range of behavioral pathologies, including rape and infanticide (Calhoun, 1962). But when salmon and trout are kept at high densities in hatcheries, aggression breaks down and peaceful schools are formed instead (Kalleberg, 1958). As a first approximation, it seems safe to assume that aggression will increase with crowding for those situations in which aggression increases access to limited resources. On the other hand, aggression imposes certain costs, in particular, expenditure of energy and the risk of possible injury. If the crowding is so great that there are simply too many individuals with whom one must compete and/or successful resource exploitation is otherwise impossible, then the best strategy might be to save time and energy and wait peacefully for things to get better.

However some correlation appears to hold between population density and social tolerance, especially among rodents. A convenient distinction can be made between *density-tolerant* and *density-intolerant* species (Eisenberg, 1967). Density-tolerant species tend to be *r*-selected. They have high reproductive rates that are maintained even as density increases; their social and reproductive systems tolerate a great deal of crowding. However dispersal occurs eventually, generally in large numbers, as in the famous case of lemmings. In contrast, density-intolerant species tend to be *K*-selected. These species generally have low reproductive rates, and reproduction often ceases altogether if the animals are crowded unnaturally, largely because aggression rises dramatically with increased density. Species of this sort tend to maintain greater year-to-year stability of population size, in contrast to the more wildly fluctuating boom or bust cycles of the density-intolerant species. Again, the extent of ag-

217

gressiveness involved in response to crowding is part of the unique adaptational syndrome of each species. (See Lidicker (1965) and Archer (1970) for accounts of different responses to crowding in different species.)

Dispersal, when it occurs, is most commonly performed by young animals, often in response to aggression from the adults, especially the adult males (Sadleir, 1965; Healey, 1967). Their own juveniles often constitute a threat to the adults' resources, and therefore aggression directed toward them is not surprising. However evolutionary theory predicts that parents distinguish between their own offspring and strangers, restraining their aggression toward the former relative to the latter. This has been confirmed for Richardson's ground squirrels, in which adult females occasionally share resources with offspring, even after the latter have left the home burrow (Michener, 1973). Among prairie dogs, parent-offspring aggression is minimal and the adults often disperse, abandoning their home area to the maturing juveniles (King, 1955).

Pikas are rabbitlike denizens of rocky slopes, generally above timberline. They remain active all winter despite severe weather, relying on dried meadow plants that they gather during the late summer and cure in what can be called hay piles. Survival through the winter depends upon having a good store of food, which in turn requires that the dispersing first-year juveniles establish their own resource supply as quickly as possible. Accordingly, these young animals assume typical adult aggressive behavior patterns very early in life, immediately at the time of dispersal (Barash, 1973b). Thus the chronology of aggressive maturation follows an adaptive regime, in contrast to the often over-simplified questions of whether aggression is present or absent.

Aggressiveness is here interpreted as a mechanism whereby each individual endeavors to get its share of appropriate resources because of the selective benefit associated with those resources. Accordingly, animals should not be aggressive when they have little or no chance of acquiring some resource, that is, when they have found themselves to be subordinate in previous aggressive competition with a particular individual. Hence, less

218

aggression should occur within stable social units. This has been found to be true of many species, for example among chickens, in which removal of the dominant individuals causes a great deal of disruptive aggression, compared with intact flocks (Guhl and Fischer, 1969). Just as subordinates are most fit if they refrain from fighting when they know they will lose, dominants are also most fit if they avoid unnecessary aggression. In fact, among some primate species in particular, dominants even tend to intervene and prevent fighting among the subordinates (Eisenberg and Kuehn, 1966; Tokuda and Jensen, 1968).

The introduction of a strange individual should also generate a substantial increase in aggression, as subordinate individuals seek to establish themselves above the newcomer and dominants seek to retain their choice position, not to mention the efforts of the stranger itself. Each newcomer also adds to overall crowding and resource competition, thus rendering aggression yet more likely in most cases. Finally, strangers are comparatively unlikely to be close relatives of the residents; therefore, altruistic restraint toward such individuals should be very low. This prediction is consistent with the evidence. Both human and nonhuman animals tend to reserve their most ferocious aggression toward strangers. Among rhesus monkeys the introduction of newcomers is several times more effective than such perturbations as crowding or food reduction in generating aggression (Southwick, 1969). Similarly, free-living house mice form small, distinct social units. Emigrants are apparently quite unsuccessful in gaining entry into these existing social units, as evidenced by the fact that there are fewer genetic differences within a resident group than between groups (Selander, 1970).

Aggressive behavior undoubtedly has certain costs associated with it, particularly expenditure of energy, expenditure of time that could otherwise be spent on something else, increased susceptibility to predation both because of preoccupation with the opponent and because of increased visibility, as well as risk of injury to oneself, and risk of injury to a relative. Of course, such behavior also confers benefits, notably access to resources, both costs and benefits being measured in terms of fitness. Animals should be selected for aggressive behavior when the benefits of such behavior exceed the costs. Again, cost-benefit analyses of this sort need not imply conscious assessment by the animals involved; selection automatically favors responsiveness to situations which maximize the ratio of benefit (in fitness) to cost (in fitness). This approach is relevant to all of the above-described circumstances that are known to influence aggression. Indeed, it would probably be a useful mental exercise for the reader to review those examples with regard to cost-benefit considerations. Cost-benefit analysis works, at least in theory. Unfortunately however, we do not yet have direct empirical measures of the consequences of aggression versus nonaggression in terms of real fitness. Perhaps some day we shall.

Animals may be induced to behave aggressively or nonaggressively or somewhere in between, by a variety of proximate routes. For an evolutionary analysis, the precise nature of these routes is unimportant, and in fact we expect a variety of mechanisms to have evolved, depending upon the species, the situation, etc. In some cases, aggression may be strongly influenced by genotype, such that appropriate stimuli "automatically" elicit inborn responses. For example, the red breast of a male stickleback fish generates aggressiveness in another male, regardless of its previous experience, so long as the context and hormonal priming are appropriate. Among mammals including humans, aggressive responses are significantly more susceptible to modification through learning and other factors of experience. Regarding human aggressiveness it may be useful to distinguish between fundamental, gut-level responses and culturally-imposed rules of expression. Thus, if we see little aggression in a courtroom, it is because the judge and bailiff impose rules of permissible behavior; nonetheless, the basic response tendencies may still be there.

The greater flexibility of aggressive response among mammals is consistent with evolutionary theory in that it is itself adaptive. Further confirmation simply requires that the proposed environmental correlates of animal aggression represent adaptive complexes; that is, there need be no real conflict between environmentalist and instinctivist interpretations of aggression, especially if it can be demonstrated that the various proximate causes of aggression themselves produce patterns suggesting that natural selection is acting to maximize the fitness of individuals. In other words, if experience A leads to aggression pattern X or, more accurately, to a possible range of patterns, X, Y, or Z, and if A → (X, Y, or Z) leads to greater fitness than does the connection A → (P, Q, and R), then the former will be an adaptive strategy and natural selection will operate to ensure its occurrence. Social scientists have largely concerned themselves with analyzing the link between A and the observed responses, the proximate causation, rather than asking why these particular linkages and not others occurred in the first place, the ultimate causation. Neither question is better; they are complementary.

Among environmentalist theories of aggression, a well-documented one is *reflexive fighting* in which animals behave aggressively in response to pain (Ulrich, 1966). Two rats caged together will begin to fight if they receive an electric shock. This correlation is almost certainly adaptive in that pain itself is often an indicator that an animal is being attacked. Aggression in both human and nonhuman animals has also been attributed to frustration, objectively defined as the inhibition of an on-going, goal-directed response (Dollard et al., 1939). Give a hungry rat apparent access to food at the end of a runway, then interpose a glass barrier just before he gets to it: The rat behaves aggressively. It is angry, and we can all relate to that. Indeed, aggression in such a case is clearly adaptive insofar as it increases the likelihood of getting one's way, especially because among free-living animals a frustrating experience is likely to be caused by another individual who is in the way.

Other environmental explanations for aggression are equally compatible with fitness considerations. It has been pointed out, for example, that many animals learn to fight by fighting, just as they learn to win by winning (Scott, 1958). Thus, a 97-pound weakling-type mouse can be transformed into a ferocious bully, and a successful one at that, by carefully pre-arranging its aggressive experiences such that it always faces opponents that are somewhat inferior. This technique is strikingly similar to that employed by managers of prize fighters who carefully bring their proteges along and do not overmatch them early in their careers, so as to develop a winning psychology. It is highly adaptive for the aggressive tendencies of an individual to be susceptible to this kind of early experience; indeed, it mimics the normal experiences of young animals as they establish their position in a social dominance hierarchy. Fitness is maximized by achieving the highest possible status role. And this is accomplished by proceeding through successive victories until a position is achieved such that one can advance no farther. In other words, most organisms are probably selected for susceptibility to victories as reinforcers of further aggression. The other extreme also holds and for equally good reasons: A constantly defeated animal is likely to be unaggressive. Such perpetual losers are to be found at the bottom of dominance hierarchies where they are most fit if they behave submissively, biding their time until they become older, larger, smarter, or stronger, and/or their superiors weaken or grow old (Plate 8.4).

Another environmentalist theory of aggression holds that animal hostility is precipitated by a breakdown of social systems

PLATE 8.4 SUBORDINATE MALE BISON *(left)* INDICATES HIS STATUS BY LOWER-
ING HIS HEAD AND TURNING AWAY FROM THE DOMINANT. (PHOTO BY D. LOTT)

(Scott, 1958). There is good evidence that this is often the case and, as discussed above, there is also every reason to expect individuals to take advantage of a loosening of social strictures by improving their access to crucial resources. Such a strategy would increase fitness and be positively selected. With this approach in mind, it becomes absurd to limit any consideration of aggression to either environmental or genetic factors acting alone. If increased fighting breaks out within a social group when the leaders are removed, this may on one level be attributed to social instability. On the other hand, to know why individuals adopt the strategy of increased aggressiveness as a response to social instability, we must ask evolution; we must inquire, what's in it for them?

In contrast to environmentalist theories of aggression, the classical European ethologists, led most notably by Lorenz, have emphasized a *dynamic instinct* concept of aggression. Although such views are ultimately grounded in evolution, i.e., the adaptive nature of aggression, they have also tended to concentrate upon proximate mechanisms, emphasizing the extent to which organisms have an innate *need* to behave aggressively. According to this approach, for example, individuals reared in a non-frustrating environment will nonetheless tend to be aggressive, and probably be intolerable brats as well. A similar argument can be used with individuals deprived of the opportunity to release aggression. The last words on this controversy have yet to be written.

Despite the occasionally aggressive debates regarding the roots of aggression, both sides can be encompassed within sociobiologic theory. In contrast, psychoanalytic interpretations of aggression do not lend themselves to a satisfying integration with evolutionary theory, perhaps in part because they have been developed almost exclusively for human behavior and, for that matter, almost always *aberrant* human behavior. Freud's *death instinct* (1920) and Fromm's *malignant aggression* (1973) have no obvious relevance to fitness maximization in that they posit tendencies that are purely destructive, with no redeeming biological consequences. The former seems closer to the old concepts of "caloric", "phlogisten", and other verbalizations that

223

substitute nomenclature for understanding, and the latter may actually represent true behavioral pathologies, unique to the complexities of modern human existence and accordingly, of little adaptive relevance. Indeed, one hallmark of sociobiologic thought is its reliance upon the essential "healthiness" of behavior; analyses in terms of fitness maximization and adaptive significance automatically assume that the behaviors in question have a positive, functional value.

Among vertebrates it is a generalization that males are more aggressive than females. Just as females tend to be more involved in parental care, with corresponding involvement of female hormones in parenting behaviors, the male propensity for aggressiveness is correlated with involvement of male hormones in aggressive behavior. Androgens, particularly testosterone, have been consistently associated with aggressiveness in a wide range of species (Conner, 1972), although the situation for human beings is considerably less clear (Ehrenkranz et al., 1974). Typically among non-human animals, elimination of male hormone by castration causes a dramatic decline in aggressiveness, followed by a return to normal levels when the missing hormone is replaced. As with female hormones and parenting behavior, the correlation between male hormones and aggressive behavior is not 100% and substantial variation occurs between species, with considerable room for environmental modification as well. As in the former case, however, the generalization is nonetheless valid, just as in both cases the identification of hormonal correlates provides a proximate mechanism.

Why are males generally more aggressive than females? Because they possess the appropriate hormones. But why do they possess those hormones? Or an equivalent question, why do those hormones have those effects? Now we are again dealing in ultimate terms and, again, sociobiology has an answer. Males are generally selected for aggressiveness relative to females because of the basic male-female differences considered earlier. Thus, because of the usual disparity between males and females regarding parental investment, males are selected for a high degree of direct male-male competition for access to females

(Plate 8.5). This in itself is a major selective pressure favoring male aggressiveness. Furthermore, since females tend to specialize in parenting, especially among mammals, males are free to specialize in defense of those resources that are crucial to fitness, for example, defense of the territory and protection against predators. Actually, males in this case are not only liberated for such activities; we might also say they are forced into this role, given the females' concentration on child care.

For most mammals in which the female and her offspring constitute the only close social unit (Eisenberg, 1966), males clearly maximize their fitness by being aggressive toward other males, whereas the benefit to female aggressiveness is much lower. On the other hand, adult females are aggressive when such behavior benefits their fitness; for example, among both grizzly bears and black bears, the adult males are a major threat, perhaps *the* major threat, to the survival of cubs. The females are accordingly very aggressive under these circumstances—in fact, this explains the well-known nasty disposition of a sow with cubs.

PLATE 8.5 DOMINANT BIGHORN SHEEP RAM INTERPOSES ITSELF BETWEEN AN ADULT EWE AND TWO SUBORDINATE MALES *(at left)*. (PHOTO BY V. GEIST)

What does evolutionary wisdom tell us about those instances when humans should be aggressive? Pathologies aside, human beings are probably selected for aggressive behavior under the following circumstances: When crucial resources are limited and obtainable by contest rather than scramble competition; when aggression leads to success, either in obtaining such resources and/or in further aggression, which of course ultimately leads to enhanced access to resources; when experiencing pain or discomfort or frustration; when social systems are disrupted and there is opportunity for advancement and/or need to defend one's situation; when strangers appear—an extreme case of the tempering of aggression in proportion to the likelihood of such behavior reducing the fitness of relatives.

It is pointless to debate whether humans are innately aggressive, especially because genetic and experiential factors are so intimately involved in the determination of such behavior. On the other hand, it may be profitable to analyze human aggressiveness in terms of its *susceptibility* to various experiences, assuming here that the pattern of responsiveness will itself be adaptive.

Certainly, one of the most powerful currents in recent human history has been the transformation of lifestyles from predominantly rural to urban. Pastoralism, nomadism, hunting-gathering, agriculturalism, these all represent fundamental patterns of human ecology. They are how 99% of us made a living, for 99% of our evolutionary history (Lee and Devoe, 1968). They also are characterized largely by scramble competition. In contrast, industrialization and the resulting urbanization provide many of the classic requirements of contest competition. Enough said?

How Far Should Aggression Go?

Because aggression imposes costs as well as providing benefits, individuals should carry their aggressiveness as far as necessary to achieve their end, but no further. This is because individuals are most fit if they maximize the difference between benefits derived from aggression and costs incurred by it. Diminishing returns may be a general case here, in that individuals profit increasingly from increases in aggressiveness but, after a certain point, further increments in aggressiveness are disadvantageous because of increasing risk to the aggressor as well as increasing danger of interfering with relatives and thus losing inclusive fitness (FIG. 8.1). This relationship between aggressiveness and cost probably holds for many aspects of aggression, including the precise circumstances under which it should occur in the first place, as well as its form and intensity.

Ethologists have emphasized that, although aggression is commonplace, it is rare for an individual to be killed by another (Lorenz, 1952; Eibl-Eibesfeldt, 1961; Tinbergen, 1968). Of

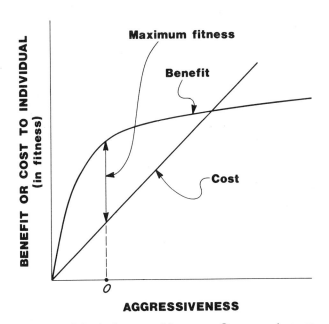

FIGURE 8.1. A GENERAL MODEL OF OPTIMUM AGGRESSIVENESS. It is assumed that the benefits associated with aggressiveness increase rapidly at first and more slowly afterwards, while costs continue to increase. Each individual will be selected for an optimum amount of aggressiveness (O) that maximizes the difference between benefits and costs.

227

course, this excludes predator-prey interactions between different species, in which killing is the name of the game. However, just as a butcher is not angry when he plies his trade with a side of beef, there is no reason to think that a lioness is aggressive when she stalks and kills an antelope. In fact, the behavior patterns employed in predator-prey encounters are generally distinct from those used when predators or prey are each interacting among their own kind. Accordingly, when biologists speak of aggression, they usually refer only to interactions between individuals of the same species.

If animal aggression rarely leads to killing, it is largely because, in proximate terms, most species have evolved elaborate and effective behaviors that enable contest competition to be resolved short of physical violence. Threat, bluff, and bluster are the order of the day. When two aggressively-motivated male fish meet at a territorial boundary, they may puff themselves up, looking as intimidating as possible. Then, assuming neither has retreated, they often orient broadside to each other and engage in a *lateral display* in which each propels a current of water against the side of the opponent. This provides each participant with accurate information about the size and strength of the other. The aggressive encounter may terminate here if either animal perceives itself as being obviously inferior. However, if they are so evenly matched that the contest is not resolved, they may face each other head on, often raising the gill covers to appear larger yet. Then, they may lock mouths and push and pull in a tug-of-war that provides further direct evidence as to the capabilities of each opponent (Plate 8.6). Most aggressive incidents are settled by the time things have reached this point, but, if not, then an actual fight may ensue. Even then, however, death is almost unheard of. The loser eventually signals submission, often by changes in color as well as by changes in behavior, and the winner generally accepts the victory without pressing his attack.

Most cases of animal aggression are thus a far cry from bloody free-for-alls. In fact, they are often so stylized as to be more aptly described as tournaments rather than as fights (Plate 8.7). The notion of animal aggression as a tournament is particularly appropriate, because it suggests that the encounter is programmed by rules and perhaps a touch of pageantry as well. Although written by evolution rather than by a human hand, the strictures appear to be no less real and adherence is

PLATE 8.6 KISSING GOURAMIS IN KISSING POSTURE. In this species, "kissing" is not a token of affection; rather, it serves as a means of assessing and maintaining dominance. (NEW YORK ZOOLOGICAL SOCIETY PHOTO)

PLATE 8.7 DOMINANCE DISPLAY AMONG GRANT'S GAZELLE. The dominant animal (*right*) performs "head flagging" toward the subordinate. (PHOTO BY F. WALTHER)

remarkably complete. Animals whose behavior is characterized by such rules are no more likely to break them than a human prizefighter is to pull out a revolver and shoot his opponent.

Certain lizards possess bony shields on the backs of their necks. During aggressive incidents they take turns ritualistically presenting their necks to be bitten by their opponent. This apparently serves to inform the animal being bitten as to the strength of the biter. Of course, the latter shows notable restraint in directing its attack only toward the part of its opponent that is best protected. More remarkable yet, cases have been observed in which a contestant has given up after *biting* its opponent without even having been *bitten* in turn. Presumably, the quitter is basing his submission on the other animal's failure to flinch after receiving his best shot, much like the little cartoon character who acknowledges defeat when the opposing giant is hit over the head with a baseball bat and doesn't even feel it.

There are two basic components to noninjurious aggression: (1) behaviors by which individuals seek to indicate and often exaggerate their aggressive competence in order to get their way without a potentially injurious fight; and (2) subordination behaviors signalling that a contestant has been defeated and that are normally respected by the victor; i.e., the battle is often not pursued to the death once the white flag of surrender has been raised (Plate 8.8). A given species may employ one or the other of these strategies or both; no species uses neither. Aggressive competition is often accomplished via vocalizations, as in birds, alligators, elk, elephant seals, or howler monkeys. Defeated individuals may simply run away, not so much a signal of defeat as a result of it. Nonetheless, they usually are not pursued by the victor, or at least not very far or with much vigor. To do so would be to risk greater loss than could be gained.

Roe deer bucks strut alongside each other during aggressive encounters in an apparent effort to present the largest possible profile. Periodically, they clash antlers and engage in a pushing contest. A buck was observed about to begin such antler clashing while the opponent was still displaying laterally, its unprotected flank exposed to the other's sharp antlers. But no attack

230

PLATE 8.8 SUBMISSIVE MALE BIGHORN RAM *(right)* SIGNALS HIS SUBMISSION
TO A DOMINANT RAM. (PHOTO BY V. GEIST)

came; the first animal simply waited until the second was ready
to meet him antler to antler (Eibl-Eibesfeldt, 1961). Of equal
significance, females that lack antlers also lack the males' inhibition against flank attacks. A defeated wolf is believed to expose
its vulnerable neck to the victor, who then refrains from killing
it, although doing so would be quite easy and would remove a
competitor (Lorenz, 1952). Even though this particular interpretation has been rather controversial, the observation
exemplifies a general trend whose validity cannot be ignored:
Many species have evolved specialized subordination rituals
that inhibit further aggression by the dominant member.

Aggression-reducing subordination rituals seem to derive
from three main sources. First, a defeated individual may signal
its defeat by turning away its eyes and mouth, thus removing a
primary source of threat (Chance, 1962); this is an alternative
explanation for the wolf behavior described above. Second,
subordination behavior may involve mimicry and/or exaggeration of some characteristic female behavior, as with *presenting*
in nonhuman primates. This association is not surprising in

231

that males often possess some adaptive inhibition against attacking females. Third, adult inhibitions against attacking young of their species are adaptive and commonly found. Accordingly, it is not surprising that adult behaviors whose apparent function is the appeasement of another adult often involve mimicry of infants. Such behavior is particularly common among courting adults, where the female must overcome male aggressiveness and/or the male must overcome female timidity. For example, the courtship ceremonies of many birds involve begging for food by the female, in which she performs a remarkable imitation of a nestling, accompanied by ritual feeding by the male, suggestive of his subsequent provisioning of offspring (Armstrong, 1958). Play and verbal infantilisms are also characteristic of much intimate behavior among adult humans. These behaviors may have similar wellsprings in terms of generating endearment in the male and reassurance in the female, although the evolved mechanism must be remote, if it exists at all.

Regardless of which proximate mechanisms are employed, the observation that animals refrain from aggression beyond a certain point requires an ultimate explanation as well. One possibility that periodically crops up is that such restraints are good for the species. Readers at this point should be able to discard this notion with some confidence. A possible but still tenuous explanation is that such behavior benefits the group and therefore arises via group selection. A better possibility is kin selection, a form of altruism. Thus, genes that induce forebearance by the victor would be selected if their effect was to save enough of the gladiators who are themselves relatives of the altruist so as to maintain or increase the frequencey of those genes, compared to hard-hearted "thumbs-down" genes that denied mercy to a defeated foe.

Perhaps the best explanation for aggressive restraint is pure selfishness of the sort described by cost-benefit analyses (FIG. 8.1). Thus, once superiority is established and along with it, priority of access to resources, nothing is gained by pursuing the dispute. Even an injured or otherwise defeated opponent could inflict a wound if it was fighting for its life, and in any case, the victor undoubtedly has more productive things to do

232

with its time and energy once its own ascendancy is assured. An instructive model for the evolution of limited war strategies among animals was published under the title "The Logic of Animal Conflict" (Maynard Smith and Price, 1973). In this study, a computer simulation evaluated individual fitnesses of various alternative strategies during aggression. Basically, this involved various combinations of conventional tactics (Marquis of Queensbury, tournament type, noninjurious behavior) and dangerous tactics (attempting to kill the opponent). Risks and benefits were associated with the various combinations, such that, for example, dangerous tactics resulted in greater benefit if successful but greater cost if unsuccessful. The results showed that selection operating upon individuals is sufficient to explain why dangerous tactics are rarely employed in aggressive contests, although a contestant was also maximally fit if it responded to its opponent's dangerous tactics by also employing such tactics. For more detailed mathematical treatments of aggression see Maynard Smith (1974), Parker (1974), Maynard Smith and Parker (1976).

Aside from the question of inhibitions against killing once aggression is already underway, limits may also be established with regard to the initiation of aggression in the first place. Thus, excessive readiness to engage in aggression, i.e., extraordinarily low thresholds, can be disadvantageous in itself. The mate and offspring of unduly pugnacious males may suffer from *aggressive neglect* if such males spend too much time hassling with their neighbors and insufficient time on parental and other maintenance duties (Ripley, 1961). This may partially explain the observation that, although heritability of aggressive behavior appears to be rather high in certain cichlid fishes, distinct low-aggression males persist within some lines (Barash, 1975e).

Finally, and at the risk of seeming to contradict some of the tidy generalizations previously stated, it should be pointed out that animal murder is in fact far more common than earlier ethological studies had suggested. Infanticide among langur monkeys has already been mentioned. Infant cannibalism has been reported for chimpanzees (Suzuki, 1971), and male

233

Japanese and pig-tailed macaques have been seen to kill other males (Kawamura, 1967; Bernstein, 1969). Hyenas, lions, and wolves behave similarly on occasion (Kruuk, 1972; Schaller, 1972) and further research will likely reveal that many animals live in something less than a state of idyllic pacificism (Plate 8.9). This finding does not violate any law of evolutionary biology, and the generalization that intraspecific killing is rare among animals nonetheless stands. The more important generalization is that behavior should be such as to maximize fitness and, significantly, in the cases of animal murder observed thus far, the victim was almost invariably a stranger having few genes in common with the killers as well as being a threat to established precedence over resources. Furthermore, the species involved were very much engaged in contest competition. Thus, the evolutionary motive is generally sufficient to make the so-called crime adaptive and thus, as far as natural selection is concerned, no crime at all.

PLATE 8.9 TWO ADULT TIGERS FIGHTING. When strangers meet, as in this case, serious injury or even death may result. (PHOTO BY J. ROGERS)

Incidentally, the above reasoning should not be taken as justifying any human behaviors on the grounds that they somehow enhance the fitness of the instigator. Evolution is a description of how the natural world works; it is not a moral end and should not be used to either justify or repudiate behavior. Extinction as well as reproductive success are both normal outcomes of the evolutionary process and therefore behaviors that further the one are. no better than those furthering the other, even if we grant the questionable assumption that behavior in support of evolution is somehow to be desired. Physicists do not claim that positive charges are good because they attract negative charges, just as there is nothing reprehensible about them either. Similarly, there is no basis for valuing courtship displays or aggressive restraint or infanticide for that matter, because they enhance the fitness of the animal in question. Remember also, these fitness increments often occur at the expense of someone else's fitness.

Evolution does not provide a natural moral yardstick with which humans can measure the actions of other animals or other humans. However we do derive our morality from somewhere and it would not be surprising if unconscious fitness considerations were the ultimate underlying source of these judgements. We may well have been selected for valuing and approving behaviors that contribute to our fitness; e.g., kindness to friends and relatives and antagonism to strangers. One problem that derives from such moral wellsprings is that optimal fitness-derived morality for me is likely to run at least somewhat counter to the same considerations for you. Perhaps the solution upon which human societies have fastened is morality imposed by societal restriction (Campbell, 1975). In any event, the point is that we are likely to misuse the natural world grossly if we consciously apply evolutionary criteria in making moral judgements. Yet at the same time, those same evolutionary criteria may ultimately be responsible for the moral judgments we unconsciously do make.

Genetically mediated inhibitions against killing other humans are certainly not well developed in our species, if indeed they occur at all. Witness My Lai, Auschwitz, Wounded Knee, the St.

Bartholomew's Day Massacre, the whole grisly history of human murder, both private and state supported. But on the other hand, although we lack the automaticity of inhibited aggression found in some species, we may not be all that bad. Chimpanzees and gorillas, our closest living relatives, are unaggressive by most vertebrate standards (Van Lawick-Goodall, 1968; Schaller, 1963), reflecting an adaptive strategy for these species in which contest competitions are relatively unimportant (Plate 8.10). Compared to many other species, our own aggressive thresholds may be rather high and indeed, an unbiased Martian observer might well conclude that we are actually rather peace loving (this image borrowed from Wilson, 1975). Such an observer might have to witness many hundreds of encounters among humans before any substantial aggression was

PLATE 8.10 A PEACEFUL SCENE AMONG THREE ADULT CHIMPANZEES AT THE GOMBE STREAM RESERVE, AFRICA. Contest competition is relatively unimportant in this species and accordingly, aggression is rare. Social grooming, as shown in this picture, is important in providing social cohesion. (PHOTO BY H. KLEIN)

apparent and probably thousands before a single injury. Our Martian zoologist might very well never witness a murder and would likely conclude ruefully, as many ethologists do when studying nonhuman animals, that *Homo sapiens* are significantly less aggressive than his own species!

What if You Win? What if You Lose?

Given that an animal has seen fit to behave aggressively and that its behavior during the encounter has been such as to maximize its fitness, what happens next? For the victor there is very little problem. It will enjoy the fruits of victory: Enhanced access to mates, food, territories, etc., all the resources that contribute ultimately to reproductive success and that made aggression adaptive in the first place. There is abundant evidence that such dominant individuals engage in more matings and hence are more fit than are subordinates (Guhl et al., 1945; De-Fries and McClearn, 1970; Geist, 1971; Schaller, 1972; Wiley, 1973; Leboeuf, 1974; Plate 8.11).

Successful competitors will assume behavior that indicates their success. This may include a characteristic way of moving, a certain regal bearing that signals dominance. For example, dominant wolves and rhesus monkeys hold their tails high and exude confidence. In contrast, subordinates often seem to slink about, making themselves appear as inconspicuous as possible. This mode of communication is highly adaptive for all concerned; a dominant animal is better off if it can gain priority of access without having to fight about it each time. Similarly, if it is going to lose anyhow, a subordinate is better off recognizing that fact and avoiding the cost associated with unsuccessful efforts at self-betterment. Selection will accordingly favor the evolutionary modification of behavior to serve communication of social status, i.e., through ritualization. And the establishment of social rankings will itself be an adaptive consequence of competitive differences between individuals (Plate 8.12).

Where dominant individuals are concerned, everything is just fine. They will attempt to maintain their advantage as long

237

PLATE 8.11 THE FRUITS OF DOMINANCE: A SUCCESSFUL BULL BISON *(left)* WITH A SEXUALLY RECEPTIVE COW. (PHOTO BY D. LOTT)

as possible, consistent with the maximization of inclusive fitness, not too much nastiness toward relatives, etc. In some cases, optimal strategy may involve extreme efforts during which the dominant animal becomes exhausted, often because of not eating, and must eventually retire, sometimes after only a few weeks as in successful harem masters among elk (Darling, 1937). Although fitness is enhanced, lifespan may actually be shortened, as in mountain sheep (Geist, 1971). On the other hand, optimum strategy for retaining dominance may require the formation of alliances with other individuals as in baboons. Here, ruling oligarchies are often composed of several adult males that owe their dominance to concerted action. Each participating male may be personally subordinate to individuals outside the ruling clique; they are dominant together but subordinate separately.

From the viewpoint of subordinates, the issues are more complex. Given that the subordinates have been bested by one or several others, why do they stay around and act as doormats?

PLATE 8.12 YORKSHIRE PIG SOW NURSING A LITTER OF PIGLETS. Dominance relationships are established among the piglets within a few hours after birth. Competition occurs for access to nipples, since the more anterior ones provide more milk. (PHOTO BY J. SPURR)

Of course, in some cases, they do not, they disperse. For other species and, sometimes, for particular individuals within a species it is preferable to remain in the social group as a subordinate rather than seek entry into another group or attempt to go it alone (Plate 8.13). Of course, "preferable" in such cases refers to "most fit." Thus, many social groups are relatively closed to penetration by outsiders, probably in large part because this itself is an adaptive strategy for most group members (Hamilton, 1975). We have already considered the numerous adaptive advantages conferred by sociality (Chapter 5). Unsuccessful social competitors may accordingly be more fit if they remain within a social network and avail themselves of its benefits, even if they have to accept subordinate status in order to do so. For example, among certain species of birds, over-wintering mortality is higher among solitary individuals than among those in flocks (Fretwell, 1969). Thus, given the unavoidable differences among individuals, competition must necessarily result in some form of ranking. If the losers are more fit by accepting that

PLATE 8.13 SOCIAL INTERACTION BETWEEN TWO MALE ARABIAN ORYX AN-
TELOPES. This species commonly forms social groups including several males,
of which one is clearly dominant. In this photo, the dominant male (*at left*) has
blocked the movement of the subordinate, who responds with a defen-
sive/submissive posture. (PHOTO BY F. WALTHER)

ranking than by refusing to participate, then some form of social dominance hierarchy will result. Furthermore, as described
above, selection operating upon both dominants and subordinates will favor efficient behavior patterns that institutionalize
these systems for the benefit of each participating individual.
This functionalist view of animal social hierarchies seems to
provide more insight than viewing the dominance structure of
each species as an immutable characteristic with a particular
pattern of empty slots to be filled by the available individuals in
each case.

Beyond the mere advantages of group living, a subordinate
may ultimately have a bright evolutionary future. Indeed, this
would have to be the case, or else acceptance of subordinate
status would be selected against. Dominant animals may constantly be tested by their underlings who are likely to improve
their positions when there is room at the top, as there eventually must be. For example, among European black grouse, predictable social increments are the prequisites of increased age.

Yearlings are relegated to peripheral breeding territories and rarely if ever mate. Two year olds enjoy better position, and three year olds may become dominant birds at the lek. These birds in turn enjoy the highest percentage of copulations (Johnsgard, 1967).

In some cases, notably in primates, subordinates often accomplish some matings; their exclusion by the dominants is not absolute. Depending upon the species and the circumstances, this may simply reflect the fact that dominant animals cannot be totally dominant all the time. Of course, subordinates will be selected for acquiescing to their subordination if this strategy confers more benefit than cost, both measured in units of inclusive fitness. Occasional lapses in the dominants' ability to dominate may therefore permit the subordinates to garner sufficient crumbs from their table to tilt the balance in favor of accepting the social system. On the other hand, the fitness of dominant individuals may itself require the cooperation of a minimum number of subordinates, so as to constitute a viable social unit. Therefore, these dominant individuals may actually be selected for permitting occasional mating success by subordinates, i.e., "intentionally" dropping some crumbs, to satisfy the fitness requirements of these subordinates and, hence, ultimately maximizing their own fitness. Certainly on a nongenetic level many human governments have recognized an analogous strategy: Allow certain essential satisfactions and/or distribute favors to retain the allegiance of their subjects.

Another fitness consideration that may predispose subordinate individuals to accept their situation may derive from kin selection (West Eberhard, 1975). Thus, if potential subordinates are sufficiently related to the dominants, an added evolutionary compensation of being subordinate could be the enhancement of the dominants' reproductive success. In the language of sociobiology: Genes that increase the likelihood of their carriers' accepting subordinate status could themselves be positively selected if they resulted in a greater net representation of similar genes in succeeding generations, because of the enhanced reproductive success of the related dominant individuals.

This success would of course have to be greater than the personal reproductive loss suffered by the subordinates. If so, then subordinates would be selected for a degree of altruistic acceptance of their lot. This would occur when $k > 1/r$, i.e., subordinates and dominants are closely related, dominants have high reproductive potential, and subordinates have a low potential. Among paper wasps, for example, adult females with the smallest ovaries are most likely to become subordinate workers during times of resource scarcity (West Eberhard, 1969). Of course, kinship considerations could also work the other way around: Dominant individuals could be selected for permitting a degree of reproductive success among subordinates, provided once again that $k > 1/r$, this time with the dominants being altruistic.

In any event, subordinate members of a social dominance hierarchy are probably making the best of a bad situation. They may eat less than the dominant individuals (Murton et al., 1966); or they may be the butt of aggression from the dominants (Greenberg, 1946). In at least one bizarre case of a unisexual species, they may be parasitized by other species. For example, unisexual species are found among minnows. Individuals of a unisexual species are all female; their eggs require stimulation by sperm from males of a closely related bisexual species, although, because fertilization does not occur, these males receive no evolutionary return. The males involved in such unrequited matings are subordinates who are reproductively excluded from females of their own, bisexual species (McKay, 1971).

Hierarchies of social dominance are nonetheless most parsimoniously interpreted as the result of natural selection operating to maximize individual fitness in the face of inequality among individuals. In some cases the resulting system may be strict and linear: A dominates B that dominates C that dominates D, etc. Rarely, it may be triangular or circular: A dominates B, B dominates C, and C dominates A. (Algebraic rules need not apply to social behavior.) In other cases, there may be frequent reversals such that the resulting hierarchy is probabilistic rather than absolute: A dominates B 54% of the

242

time they interact, B dominates A 46%, etc. Characteristics that contribute to dominance appear to include size, strength, quickness, being related to someone who is already dominant, luck in landing blows (Collias, 1943), and in some cases even intelligence (Van Lawick-Goodall, 1971). As observers, we assess the relative dominance of animals by such criteria as the outcome of fights, who pecks whom, who threatens whom, and who gains direct access to such contested resources as water, food, mates, or nesting places. Often the indicators of dominance are subtle, such as posture or passive avoidance of the dominant by the subordinate. Substantial familiarity with the animals may therefore be necessary in order for any social ranking to be apparent. This itself is testimony to the efficiency with which many dominance hierarchies operate.

The concept of social dominance has often been used in analyzing the structure of animal social systems, particularly nonhuman primates. Recently however there has been increasing recognition that dominance may have been overvalued as an organizing principle and that it should be replaced by the concept of social role (Gartlan, 1968). It may be more profitable, to identify roles relating to the initiation of group movements, defense against predators, attempting new foods, interacting with other groups, etc. Individuals may assume different roles for each situation rather than being type-cast into a unitary framework of dominance ranking. Certainly, role orientation and/or dominance status change during the lifetime of each individual. Furthermore, recent studies of African black and white colobus monkeys suggest that groups themselves may undergo changes in basic social structure as the individuals comprising them grow older and assume new social relationships (Dunbar and Dunbar, 1976).

Some mention should also be made of the hormonal consequences of social competition. Subordinate individuals, especially males, may experience a dramatic reduction in gonad size and hormonal output. This *psychological castration* has been described for nonhuman primates and is apparently an effect rather than a cause of low social status. Accordingly, it can be reversed if social status is artificially enhanced, for example by

243

removing the dominants, thereby permitting the former subordinate to assume dominant rank. Given the association between gonadal hormones and aggression, psychological castration may be one proximate mechanism whereby subordinates physiologically ensure the peaceful acceptance of their subordination. Significantly, they retain the potential for normal reproductive function, should this eventually become feasible.

Another hormonal correlate of social competition is even more problematic. Subordinates frequently have enlarged adrenal glands that in turn are associated with renal failure, impaired antibody formation, lowered fertility, and increased susceptibility to disease. Death may even result. Significantly, a similar and possibly identical syndrome is often associated with increased population density (Christian and Davis, 1964; Christian, 1968). High population density may subject individuals to a high frequency of aggressive encounters and, hence, it is not surprising that it may mimic subordination in its physiological effects. The problem is: Why do these effects occur in the first place? What is their adaptive significance? These phenomena may be a laboratory artifact and indicate such abnormal conditons that they could not be considered products of natural selection. Alternatively they may involve a form of altruistic self-sacrifice insofar as the reproductive success of the surviving animals is enhanced.

At this point, a basic caveat is in order. The underlying premise of this book is that animal social behavior can profitably be viewed as a consequence of natural selection. However, it does not necessarily follow that every identifiable phenotype must therefore be adaptive and the product of positive selection. For example, flying fish glide through the air to escape their predators. They possess numerous adaptations, such as enlarged, winglike fins, that are clearly the result of natural selection acting to maximize the fitness of the individuals adopting this strategy. At the end of their aerial glide, they fall back into the water. This is a simple consequence of gravity, and clearly it would be absurd to consider it further evidence of nature's handiwork (Williams, 1966a). On the other hand, if one consequence of falling back into the water was a high frequency of ruptured organs or internal hemorrhaging, selection would doubtless favor either the elaboration of protective devices on the bellies of flying fish or a different way of evading predators!

Back to aggression, population density, and enlarged adrenal glands. Animals undergo a characteristic response to stress, and aggression is a potent stressor: the pituitary gland releases a hormone that in turn activates the adrenal glands to secrete their hormones, specifically corticosteroids. These hormones apparently help resist stress (Selye, 1956). Thus far, the system is adaptive. However, prolonged stress may produce pathological results: the adrenal glands enlarge as they strain to accommodate the excessive demands made upon them. The eventual outcome is pathological, the consequences of pushing an otherwise adaptive system beyond its capabilities. The role of evolutionary processes in such a case may be indirect at best and just as with the flying fish, we should beware of seeing adaptive significance in every quirk of nature.

Finally, what about *Homo sapiens* in terms of the social consequences of aggressive competition? The vast majority of human societies were and still are polygynous, with the number of wives per husband generally correlating with the latter's social status (van den Berghe, 1975). It is therefore almost inevitable that through most of human history the variance of male reproductive success has been greater than the variance of female reproductive success; i.e., some males have produced a disproportionately large share of offspring while others have been exceptionally unsuccessful. It also seems reasonable that reproductive success has correlated with some phenotypes that confer social status, such as intelligence or physical strength. Insofar as these factors are genetically influenced, the biological characteristics of our species may be in part a consequence of our own social behavior. Of course, we cannot deny the possible role of such nonbiological factors as luck or inheritance of goods.

Certainly, optimum functioning within a social system requires careful judgment about one's social relationship relative to another, including estimation of the most propitious time for mounting a challenge against someone of higher rank. The substantial benefits and costs associated with making these decisions may even have contributed to the rapid evolution of intelligence among primates (Chance and Mead, 1953). However, despite our vaunted intellects and our protestations of personal freedom and independence, humans show a disturbing tendency to accept subordinate roles, even when this involves doing serious harm to another (see, for example, Milgram, 1974). The inclusive fitness of individual proto-hominids within prehistoric hunting-gathering bands may have depended heavily upon a willingness to accept a degree of subordination within the given social unit.

Yet at the same time as we may have been selected for a degree of subordination and indoctrinability, correlations of status with fitness may also have selected for efforts at self-betterment, whenever this is possible. The psychiatric theorist Adler follows this theme: "Whatever name we give it, we shall always find in human beings this great line of activity—the struggle to rise from an inferior position to a superior position, from defeat to victory, from below to above" (1932). Adler was not an evolutionist, but he may nonetheless have seen the footprint of natural selection operating through social competition.

Within present-day human families, dominance is almost invariably sexist, males over females, and ageist, older over younger, except for a diminution with extreme old age. This holds cross-culturally as well (van den Berghe, 1975). These patterns may or may not be adaptive; they may or may not be at least partially a product of our biology; and we may or may not choose to call them good. But they are strikingly similar to the results of social competition within most other vertebrates. And at least to some extent, they may have arisen for similar reasons.

N I N E

Strategies of
Spatial Competition:
Territories and Such

All living things have certain requirements for survival and reproduction. Imagine that animals are somehow distributed
across an environmental landscape that provides these diverse
requirements in patterns that vary in space, time, and quantity.
Given that each individual will be selected for behavioral
strategies that maximize its fitness, a variety of social groupings,
mating behaviors and parenting behaviors will result, tailor
made by natural selection to balance the potentials and the requirements of each species in each case. Furthermore, if individuals are more fit contesting for resources than scrambling
for them, then various strategies of competition will also evolve.
Social competition will be mediated by agonistic behavior; if the
nature of the species is such that individuals are more fit by remaining in social units, then some sort of dominance hierarchy
will often result (Chapter 8). Competition therefore produces
ranking when resources must be shared among individuals that
are inherently unequal.

On the other hand, the nature of the resources combined
with the nature of the species also specifies a certain optimal
pattern of physical dispersion in each case. Thus, all animals
distribute themselves spatially in some way or other, and this
chapter will consider the strategic aspects of these various techniques of dividing space among individuals.

247

How Can Space Be Divided?

Most animals distribute themselves in space via both long-distance movements and short-distance adjustments. The long-distance movements usually occur only once during an individual's lifetime, when it disperses from the area and/or social group of its birth, eventually establishing itself somewhere else or perishing in the attempt. Of course some animals do not disperse. Eventually they simply become incorporated into their original social network; this is especially true of female ungulates, social carnivores, and group-living primates. When dispersal occurs, it often has the effect of keeping down the local population density (Lidicker, 1962), although evolutionary theory suggests that population effects may be a *consequence* of dispersal rather than a *cause*. Thus, dispersal appears to represent a strategy in which certain individuals indulge; they do so when it is likely to enhance their inclusive fitness via their personal reproductive success and/or that of kin left behind. Only when the latter consideration is important would reduction in population density be an ultimate driving force for dispersal itself.

Dispersal can generally be explained as a spatial strategy maximizing individual fitness. It is performed (1) by young animals who have not yet established themselves and who are less likely to do so than are adults; (2) by males who are more likely to be disadvantaged by reproductive competition with resident males than females are by resident females; (3) in *r*-selected species, with a syndrome of adaptations promoting success as colonizers, and (4) at times of resource stress, such as that caused by high population density and/or shortage of food, cover, nest sites, etc. These factors all make it likely that dispersing individuals are following a strategy of personal fitness maximization. Of course it is also conceivable that group selection is involved, preventing overpopulation within groups and thus contributing to group success, and/or kin selection is operating, contributing to the reproductive success of relatives who profit from the reduced competition once the dispersers have altruistically departed.

In any event, dispersal helps establish the geographic distribution pattern of a species. Whether or not dispersal has occurred in any particular case, short-distance settling movements also take place, and these establish the local pattern of who uses what space. This will be our primary concern. However, it should be mentioned in passing that an intermediate class of spatial adjustments also occurs: migration. This involves regular movements, generally between two distinct regions. These movements tend to be seasonal, connecting areas of occupation which enhance the fitness of the occupants in each case. For example, large hoofed mammals such as caribou and elk often migrate between summer and winter ranges such that they always occupy environments where the climate and food conditions are optimal. The distances travelled may be very long; for example, golden plovers migrate annually from the northern Arctic to the southern Arctic. However, for our purposes migratory movements are conceptually intermediate between dispersal and settling movements, because they simply connect areas within each of which a characteristic spacing pattern is found. For example, many migratory songbirds defend territories during the breeding season in North America, then migrate south to the tropics where they spend the summer in winter flocks.

When animals are no longer in transit, they establish characteristic spacing patterns. Basically, there are only three ways in which objects can distribute themselves across a fixed area: random, clumped, or regular (FIG. 9.1). Of these, random spacing is the rarest among vertebrates, although it has been suggested in some cases, as in certain groups of kangaroos (Caughley, 1964). The other two patterns, clumped and regular, suggest either nonrandom organization of the environment or nonrandom patterns of attraction and repulsion among the individuals themselves. Clearly, both occur. We have already discussed some aspects of social attraction (Chapter 5); environments themselves may be relatively homogeneous, as in grasslands, or clumped as with fruiting trees for chimpanzees

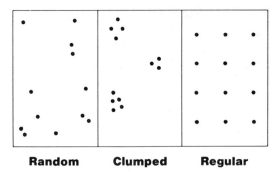

Random   Clumped   Regular

FIGURE 9.1. THE THREE BASIC PATTERNS OF
SPATIAL DISPERSION.

or a moose for wolves. Certainly, the distribution of environmental resources exerts strong selective pressures upon the optimum spatial distribution of animals: clumped resources → clumped animals; dispersed resources → dispersed animals (Horn, 1968; C. Smith, 1968). The dots making up FIG. 9.1 can be considered as either individuals or groups, depending upon which constitutes the basic social unit of the species, i.e., the smallest social unit in which members survive and reproduce normally (Brown, 1975; Brown and Orians, 1970).

A noted ethologist once told the following story.*

> One very cold night a group of porcupines were huddled together for warmth. However, their spines made proximity uncomfortable, so they moved apart again and got cold. After shuffling repeatedly in and out, they eventually found a distance at which they could still be comfortably warm without getting pricked. This distance they henceforth called decency and good manners.

The porcupines' solution to the problem of optimal spacing provides a convenient allegory. Extremes of distance or closeness tend to be disadvantageous just as extremes of aggression, or reproductive effort, or in fact virtually anything, and so the optimal phenotype is a compromise, such that maximum benefit is realized at minimum cost. In other words, natural

---

*Paul Leyhausen; quoted in Wilson, (1975).

selection will favor spacing patterns that maximize individual inclusive fitness.

All animals maintain characteristic *individual distances* between themselves and their fellows. This is readily seen by looking at swallows on a wire, gulls on a beach, or, for that matter, people on park benches. If another member of the same species encroaches within the three-dimensional personal space surrounding an individual, the latter responds with agonistic behavior, retreat, or some indication of discomfort. Some *contact species* such as hippopotamuses and bonnet macaques maintain essentially no individual distance much of the time. Other, *distance species*, keep their fellows at arm's length; as in the case of swallows and pig-tail macaques (Hedigger, 1955; Conder, 1949).

The adaptive significance of contact versus distance strategies has not yet been investigated. There may be considerable individual variation in personal space, often with males maintaining greater distance than females; this is consistent with the males' greater aggressiveness (Marler, 1956). Personal space among *Homo sapiens* varies dramatically among different cultures, such that persons from Mediterranean and Latin American societies tend to maintain considerably less space than do North Americans and Western Europeans (Hall, 1966). As in other animals, human spatial requirements also vary with the situation, decreasing as crowding increases and increasing among individuals of higher social rank (Barash, 1973c). Of course, requirements of personal space vary with other circumstances: contact during copulation; varying patterns of spacing while feeding, depending for example upon the likelihood of interference, assistance, etc.; more space when giving birth and hence more liable to predation, and so on. The adaptive aspects of these variations should be obvious, as well as the presumed mechanism whereby natural selection has favored the appropriate pattern of responsiveness in each case.

In addition to an individual's private spatial envelope, each individual in pursuit of its daily activities traverses a certain area, its *home range*. An individual's home range is often quite restricted, sometimes surprisingly so. "Free as a bird" is a mis-

251

nomer, for although most birds possess the physical equipment to go almost anywhere, most also possess certain internal inhibitions against doing so. The same applies to nearly all other animals. Chase a deer long enough and it will eventually double back, restricting its movements to a limited area, often less than 100 acres. Trap a field mouse and release it; unless it is in the process of dispersing, it can probably be recaptured again and again, almost always within a radius of about 10 yards from where it was caught the first time. Baboon troops restrict their movements to relatively circumscribed areas within which all necessary resources are to be found (Altmann and Altmann, 1970). Consider your own life: For reasons of habit perhaps or the constraints of complex society, humans do much the same. Increased efficiency generally makes this pattern appropriate for us (adaptive?), and such patterns of restricted area may act similarly upon other animals as well.

Home range is defined simply by the *fact* of use, not by *how* the space is used. Thus, overlap with other individuals may or may not occur. Monkeys within a troop may share the same home range that may itself overlap with the home range of neighboring troops. Home ranges may vary with the age of an individual (Ralph and Pearson, 1971; Fitch and Shirer, 1970), often increasing as the animal grows older. Males generally have larger home ranges than females (e.g., Brown, 1966), and the typical pattern among mammals is for the home range of males to overlap that of several females (Eisenberg, 1966), although this is by no means universal. Home range often varies with the stage of the breeding cycle (e.g., Weeden, 1965; Stefanski, 1967), but the pattern may vary from a large initial range that becomes progressively more restricted as breeding proceeds to a home range that is small initially but that expands over time.

These variations, while seemingly chaotic and resistant to generalization, are in fact susceptible to one overwhelming generality: They represent adaptive variation, appropriate for the maximization of fitness in each idiosyncratic case. Thus, home ranges that are maintained or even expanded in size during the breeding season are those in which feeding takes place, requiring a large area for the provisioning of young. In contrast, among some species, e.g., the green heron (Meyerriecks, 1960), a large display area is maintained initially by the male; the area gradually shrinks to the region immediately surrounding the nest itself, because feeding occurs away from the breeding area.

Animals must satisfy all their biological needs within their home range. Since food is one of an animal's most easily measured requirements, considerable research has been directed toward correlating size of home range and energy needs, and the results have been substantial as well as consistent with notions of fitness maximization. Accordingly, size of home range generally increases predictably with body weight in mammals (McNab, 1963), birds (Armstrong, 1965), and, to some extent, lizards (Turner et al., 1969). Furthermore, size of home range increases more rapidly for predators than for vegetarians (Schoener, 1968); i.e., the difference in size of home range between a large-bodied plant eater and a small-bodied plant eater is not as great as that between a large-bodied animal eater and a small-bodied animal eater. This agrees with ecological theory in that predators tend to be less abundant than their prey and therefore require more space in order to survive. It may take 100 acres to support ten deer, and 100 acres may support only one cougar. Correlations of size of home range with energy requirements also accord with sociobiologic theory in that maximization of net benefits would require that individuals utilize a home range that is as large as necessary but not too large. And this seems to be what happens.

Although home ranges may overlap significantly, there is often within each home range a relatively small region that is less likely to be shared and within which the residents concentrate most of their activity. Adult woodchucks are usually not found in another adult woodchuck's burrow and they are most likely to be found near their own. Coati-mundis, tropical relatives of raccoons, concentrate their movements within a restricted subset of their total home range (Kaufman, 1962), and baboons rarely wander far from sleeping trees and/or sources of water (Kummer, 1968; 1971). These *core areas* usually include a critical resource such that proximity to it is advantageous.

There is a continuum between home ranges with variable overlap and those in which overlap is prevented by agonistic behavior. The latter are considered *territories*, for which a good working definition is "a fixed area from which intruders are excluded by some combination of advertisement, threat, and attack" (Brown, 1975). Agonistic behavior is thus closely interwoven with territoriality. This should not be surprising, given that territorial behavior is a function of competition among individuals. The connection is closer yet because territory owners can be viewed as individuals that are dominant so long as they remain in a certain place. This *site-dependent dominance* is demonstrated by the fact that the proportion of agonistic interactions that an individual wins often declines as that individual goes further from the center of its territory (Brown, 1963).

In practice, it is often difficult to identify the proximate causes of nonoverlap, and in some cases agonistic behavior may rarely be seen. This may in part be attributable to the remarkable respect paid by neighbors to prior success in agonistic territorial defense by an adjacent proprietor. In addition, scent markings or other passive means of defense may be employed, and these are often difficult for a human observer to detect. Black rhinoceroses have been considered nonterritorial because little agonistic behavior occurs at boundaries, yet spatial overlap is minimal (Schenkel, 1966). Therefore, a functional definition of territories has been proposed (Pitelka, 1959): territories are defined as "areas of exclusive use," without recourse to how this exclusion is achieved. Although this approach seems to emphasize ultimate function more than does the concept of defended area, the latter seems intrinsic to most intuitive conceptions of territory; it remains the one most commonly employed in sociobiology and will be used here as well, albeit reluctantly. Nonetheless, exclusive or at least restricted use remains the *consequence* of territorial behavior, with aggression, advertisement, or display the mechanism whereby such use is achieved.

There is a truly bewildering array of different patterns of

territorial behavior, varying in the type and intensity of defense, the daily and/or seasonal changes, the pattern of who does the defending and against whom. A review of this enormous topic will not be attempted here; rather, certain generalizations will be described. For more details, see Carpenter (1958), Brown (1964 and 1969), Ewer (1968), and Bates (1970). We can identify several basic kinds of territories, in terms of what is defended:

1. Large, multipurpose territories in which mating, rearing young, and feeding take place; e.g., typical songbird territories, a classic case being the song sparrow (Nice, 1937, 1943).
2. Territories used for rearing young only; e.g., most birds that breed in colonial groups (Crook, 1965).
3. Territories used for mating only; e.g., most species using leks (Snow, 1963; Leuthold, 1966; Wiley, 1973).

In lesser known cases, individuals may defend feeding areas, roosting places, or space surrounding a moving herd of females.

In the next section we will consider some of the strategic, adaptive issues surrounding the decision whether or not to defend a territory. For simplicity it will be necessary to assume that territoriality is a yes-no proposition; i.e., either it occurs or it does not. Given the complexities indicated here, it should be obvious that this is an oversimplification. Popular treatments of territoriality such as *The Territorial Imperative* (Ardrey, 1966) have created the illusion that territory is universal among animals and, further, that humans are necessarily territorial as well. A subsequent and equally fervent effort maintaining that dominance is also universal among living things appeared in *The Social Contract* (Ardrey, 1970), thereby perhaps diluting the cogency of the earlier argument. In fact, both social dominance and territoriality are by no means universal. Their presence and absence are generally predictable (and certainly interpretable) by considerations of fitness.

In any case, there has been much (perhaps too much) impassioned discussion as to whether *Homo sapiens* is territorial (Montagu, 1968). Perhaps the best answer at this point is simply, we do not know, or, better yet, it depends upon the definition one employs. Certainly, humans do defend physical space on some occasions, perhaps the best examples coming from the cross-cultural tendency of human families to identify some form of living space and the almost universal tendency of intruders to await some form of invitation from the proprietors and/or to pacify them with a gift, word of greeting, or some request for permission or a token of subservience or at least, of nonaggressive intent. On the other hand, there is often little if any actual aggression or even exclusive use. Advertisement and display are quite variable, and not everyone agrees that good fences make good neighbors. Certainly, as a species we respond aggressively to foreign invasions of our home turf and are often difficult to defeat under such circumstances; witness the extraordinary successes of anticolonialist wars of liberation. However, such actions usually derive at least in part, if not entirely, from strong governmentally generated social pressures. We definitely have home ranges; by the way they are defined, they cannot be avoided. However as far as territories are concerned, it really does depend upon our definitions and which evidence we choose to consider. Clearly we are complex creatures, not easily classified.

A final word about words: As symboling creatures, words are essential to us; we organize our sense impressions of the world around us and our ideas about the world via our words. But words can also be traps. The natural world has not been made to fit discrete classifications but, rather, the other way around. Thus, we should not be surprised or upset if nature does not conform to our penchant for descriptive clarity and analytic tractability. We should beware the tendency to behave like the mythical character Procrustes who shortened or lengthened his victims, whichever was necessary in each case, to make them fit his iron bed of nails. Living things are what they are, and they do what they do, in many cases because natural selection acts upon each particular and unique constellation of characters to produce a phenotype that is maximally fit in each case. Indeed, we may not have enough words to describe each unique, adaptive result and, even if we did, it probably wouldn't make any difference to the varied products of evolution. So, we must use words, but we must also be careful that words do not use us.

Should You Defend a Territory?

Where species are concerned, scramble competition is more efficient than contest competition; a given quantity of resources will support a larger population of scramblers than competitors (MacArthur, 1972). This is essentially because time and energy that is expended in agonistic behavior could otherwise be spent on more directly productive endeavors. For example, defense of a territory carries with it no intrinsic benefit to the individual, as compared to foraging, mating, provisioning young, etc. So why does territorial behavior occur at all; under what circumstances would it be adaptive? Remember, natural selection acts upon individuals, not species. Territoriality can enhance the fitness of an *individual* if it ensures access to resources that ultimately contribute more to that individual's inclusive reproductive success than it detracts. Thus, territorial behavior could evolve even if it resulted in a *lower* species-wide population than did the absence of territoriality, so long as it enhanced the fitness of each individual. For its part, territorial defense can actually be relatively efficient provided it does not require constant repetition. In other words, it is worthwhile for individuals to spend time in defense of a territory, as long as they will not have to repeat the performance with equal frequency and intensity each time.

If boundaries, once they are established, can be maintained by a minimum expenditure of additional time and effort, then the proprietor of a territory is free to reap the benefits of access to resources within the territory. This is the way it usually works: The major activity in defending a territory occurs at the time of establishment, after which boundaries tend to be respected and, in fact, territorial neighbors apparently come to recognize each other (Weeden and Falls, 1959). All this may ameliorate the inefficiencies of territorial systems relative to uncontested scrambles, but the fact remains that something is nonetheless wasted in the process. In this sense, the occurrence of territoriality itself supports an important evolutionary principle: Selection operates upon individuals rather than species or populations. Thus, each individual is not concerned that a maximum number of its colleagues be supported on a given habitat. Rather, each is out to make the best evolutionary deal it can for itself, and accordingly it will be selected for strategies such as territoriality that enhances personal fitness, even though the result *may* be suboptimal where the entire species is concerned.

But not all animals defend territories. Who should? And when? Once again, individuals should defend territories when the benefits in doing so exceed the costs in fitness for each individual. Over the years, numerous suggestions have been advanced for the adaptive value of territories; among the most popular have been guaranteeing a food supply, defending the nest or den and young, facilitating pair formation, reducing disease transmission, preventing interference with copulation, maintaining the pair-bond, and limiting population density (Hinde, 1956). Most of these presumed benefits can be achieved without territorial behavior per se. The following section will consider what is probably the most compelling and parsimonious sociobiological theory of territorial behavior.

This theory suggests that territoriality is adaptive when two major requirements are met. (1) There must be competition among individuals. There is no reason for defending a resource unless someone else contests it with you. (2) The resource must be economically defendable (Brown, 1964). When both conditons are met, then the benefits are sufficient to compensate the costs of agonistic behavior, and the latter is therefore selected. Accordingly it is not economical for baboons that feed primarily on grasses to expend energy defending individual patches of lawn. Imagine an individual who does so: Because the grasses are distributed widely and homogeneously, other individuals can simply move aside and continue to feed just as well as their pugnacious colleague, in fact better, because the defending animal will be wasting time and energy while nonterritorial individuals are gathering an equal number of calories without the wasted expenditure. Thus, individual baboons who adopt a territorial strategy under such circumstances will be selected against.

Similarly, resources may fail the test of economic defendability if they are variable in location. We have already mentioned the absense of feeding territories among oceanic birds. Characteristics of the species itself may also predispose either toward or against territoriality, in that defense of an area is feasible only if one has the physical equipment to defend it. For an extreme case, sedentary oysters and barnacles are ill equipped to patrol a territory, recognize intruders, and oust them if necessary. Of course these species would probably not profit from territorial behavior anyhow, given that their main requirement is a supply of oceanic plankton, a resource that, like air, is not itself economically defendable. If it was, selection would likely have equipped these filter feeders with the physical characteristics necessary to behave territorially.

Territorial behavior is probably most characteristic of birds and certainly it has been most carefully documented in that group (see Howard, 1927, for a very readable and pioneering study). In contrast, home ranges are more typical of mammals (Burt, 1943), although some degree of territorial defense has been increasingly recognized in mammals as well, as field studies become more detailed and of longer duration (Plate 9.1). However, where some distinction remains, it may derive in part from the higher metabolic rate of birds that in turn means their resource needs are more intense, thus rendering competition more likely. Furthermore, because birds have excellent vision and can fly, they are physically adapted, one might say pre-adapted, to monitor a piece of turf for trespassers and to respond to their transgressions. Mammals on the other hand are less well equipped for such activities, although they do tend to have an excellent sense of smell and often make substantial use of chemicals, *pheromones*, to mark spatial boundaries.

There is overwhelming support for the sociobiologic theory that territories will be defended when it is adaptive for indi-

PLATE 9.1 A TERRITORIAL MALE GRANT'S GAZELLE *(right)* PREVENTS A FEMALE FROM LEAVING HIS TERRITORY BY BLOCKING HER PATH IN THE "BROADSIDE" POSITION CHARACTERISTIC OF THIS SPECIES. Territories are valuable insofar as they confer access to resources that contribute to fitness. In this regard of course, females are a resource of particular value and it is adaptive for males to keep them within their territories. (PHOTO BY F. WALTHER)

viduals to do so, i.e., when their personal inclusive fitness is maximized. According to theory, territories should be defended if such a strategy confers net advantages that would not otherwise be available. The Galapagos marine iguanas are the world's only ocean-going lizards. They feed on marine kelp; this food is incredibly abundant and no advantage would accrue to defending feeding territories, so none is defended. Nest sites are similarly abundant and are also undefended . . . on all the Galapagos Islands except one. On Hood Island, nest sites are rare and, hence, as expected, females there are brightly colored and engage in malelike agonistic bouts in defense of this scarce and crucial resource (Eibl-Eibesfeldt, 1966). In this particular environment, competition and economic defendability conferred a strong selective advantage to each individual that successfully defended a particular type of territory. Wildebeests graze the African plains; it would not be adaptive for them to defend feeding territories, and they do not do so. On the other

PLATE 9.2 TWO MALE TOPIS CHALLENGING EACH OTHER IN A BATTLE OVER THE POSSESSION OF MATING TERRITORIES. (PHOTO BY F. WALTHER)

hand, adult males do defend small territories, leks, that mediate male-male competition and male-female attraction and serve for breeding only (Estes, 1974; Plate 9.2).

Once territories are defended, theory predicts that they should be just large enough but not too large, i.e., an optimal mix of benefits and costs. Tree squirrel territories are just sufficient in size to provide adequate food for their proprietors (C. Smith, 1968). Ovenbirds are small inhabitants of deciduous woodlands (oak, maple, hickory, etc.), where they forage for insects among the leaf litter on the forest floor. When the insect population is low, territories are large, thus assuring enough food for each mated pair and their offspring. When food is abundant, territories contract (Stenger, 1958). Apparently, if sufficient resources can be guaranteed within a smaller territory, then it is maladaptive to defend an unnecessarily large one. Selection will therefore favor flexibility in territory size, always maintaining the benefit-cost advantages. Adaptive variations have been described in different species of chipmunks, in which the presence or absence of territoriality is related to the costs of such behavior, particularly the energetic expense of running about in hot, arid environments, as well as the risk of predation (Heller, 1971).

Much of the evidence that territory-holding represents an adaptive strategy comes from the two types of evidence presented thus far: qualitative adjustments in the presence or absence of territories, e.g., wildebeest and Galapagos lizards, and quantitative adjustments in some aspects of a territory such as its size, e.g., ovenbirds and predatory versus herbivorous species. There is further evidence. Under conditions of crowding, many species revert from territorial behavior to dominance structuring (Barnett, 1958; Bronson, 1963; Kummer, 1971). Proximately, this may occur simply because the crush of too many individuals makes it impossible to maintain the integrity of territorial boundaries. Ultimately, it is adaptive in that when competition is too great, resources are no longer economically defendable.

The other side of the coin from "who should defend territories" is "who shouldn't?" The answers are fairly clear, and the results are consistent with expectation. Individuals should not defend territories if their inclusive fitness will not benefit from such behavior. Accordingly, subordinates should be less concerned with territorial defense than should dominants, since subordinates are less likely to have offspring within the group and, in any case, they have less riding on the disposition of the resources at issue. For similar reasons, young animals should be less territorial than adults. Finally, females should generally be less territorial than males and, when territoriality occurs in females, it should be directed primarily against other females, just as male territoriality should be directed primarily against other males. There are exceptions, but in general this is indeed the way living things operate.

In addition, increased competition from other species with similar resource demands should have an effect similar to reducing the density of a resource. Thus, the average territory size maintained by song sparrows increases as the number of other competing species increases (Yeaton and Cody, 1974). Theory also predicts that territorial behavior would be directed particularly toward other individuals of the same species, because they constitute the greatest competitive threat to any would-be proprietor. And this is certainly true. It makes little difference to a golden eagle if a pair of sparrows nests nearby; eagles eat small mammals while sparrows eat seeds. On the other hand, the sociobiological theory of fitness maximization predicts that territorial behavior should be increasingly likely among species that are increasingly similar in their ecological requirements (Orians and Willson, 1964). This is generally the case; for example, marsh-dwelling red-winged blackbirds and yellow-headed blackbirds exclude each other from their breeding territories, whereas neither species bats an eye at a mallard duck or great blue heron who elects to share their turf. On a proximate level, interspecific territoriality may well be more likely among closely related species because such species share common physical features, *releasers*, that stimulate agonistic behavior. On the level of ultimate causation, responsiveness to the releasers of any competing individual would be selected.

In some cases, species that overlap spatially and that compete for resources have evolved so as to become more similar to each other. This is known as *convergence*. It apparently facilitates mutual recognition by the competitors involved (Cody, 1969). The usual occurrence, however, is for competing species to become more different in areas where they overlap. This is called *character displacement* (Brown and Wilson, 1956), and it appears to be a straightforward case of directional selection working in opposite directions upon individuals of two species: Consider two species, A and B, found in the same place and consisting of individuals that overlap with regard to some character related to resource utilization, for example, bill size, that in turn correlates with size of seed that is eaten. Individuals of species A whose bills are the same size as some members of species B will be competing with those B individuals, and the same applies to B individuals competing with similar individuals of species A. In contrast, individuals of species A whose bill size does not overlap that of species B will suffer less competition and accordingly, will leave more offspring; they are more fit and are selected. The same applies to the nonoverlapping individuals of

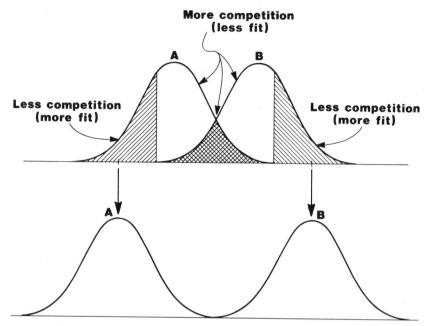

FIGURE 9.2. THE EVOLUTION OF CHARACTER DISPLACEMENT. If the distribution of some trait related to resource acquisition, e.g., bill size, within one species overlaps a similar trait in another species, natural selection will cause these species to become more different wherever they occur together. In the case shown here, individuals at one extreme of the distribution within species A are more fit because they compete less with members of species B; the same applies to individuals at the other extreme within species B. Directional selection therefore operates in different directions within the two species, causing them to diverge.

species B. Species A and B will therefore diverge, at least insofar as bill size is concerned. This evolutionary change will occur because of the greater fitness of the more different individuals of each species (FIG. 9.2).

Character displacement is an important evolutionary process, and it is generally believed to have been involved in the multiplication and increased differentiation of species: Two similar populations become separated geographically as a result of which they develop somewhat different distributions of characteristics but still with significant overlap. When they eventually come together again, character displacement occurs, thus distinguishing the two populations even more (Mayr, 1970). Biologists have emphasized almost exclusively the role of ecological competition in generating character displacement. However, social competition may also be involved. Thus, socially dominant species tend to have wider niches, a wider range of resource exploitation, than do socially subordinate species, which often must make do with narrower niches (Morse, 1974). See Murray (1971) for an analysis of the ecological consequences of interspecific competition.

Competition between species may also have been important in human evolution. Thus, much of our sociality may have evolved as an adaptation to achieve social dominance over the large carnivores with which our ancestors competed on the African savannah. We may also have competed with other primate species, both before and after we forsook the forests for the savannahs. Although most anthropologists feel that our hegira was necessitated by conditions of drought, it is also possible that we were kicked out of the trees by other primate species whose ecological or social competition forced such a strategic retreat. Alternatively, it has also been suggested that we exerted similar pressure on other primates, most notably chimpanzees that are therefore perceived as a degenerate form, banished back into the forests by our pushy ancestors (Kortlandt, 1972). However, this is a minority viewpoint.

We have considered territorial behavior as though it is only the prerogative of individuals. Of course territories may also be defended by groups, although when this occurs it may still be seen as an adaptive strategy employed by each participating individual within such groups. Larger groups are generally more successful in territorial defense than are smaller ones, and this in itself may predispose toward group living as well as group defense. On the other hand, the defended resource(s) must be sufficient to allow for individual fitness considerations, given the need for sharing within the group. Furthermore, this is probably a major constraint upon the permissible upper limit of group size. Given that larger groups do better at defending territories, why don't they get bigger and bigger, in a never-ending upward spiral? The answer is undoubtedly that beyond

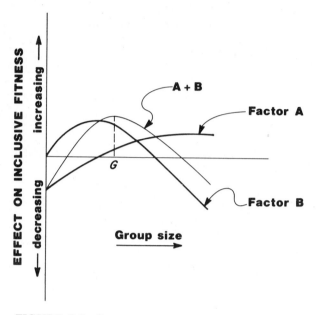

FIGURE 9.3. OPTIMIZATION OF GROUP SIZE. The optimum group size (G) is the number of individuals for which the sum of social contributors to fitness is maximum. In this simple case, only two factors, A and B, are shown. Factor A could relate to success in group defense of a territory where larger group size leads to greater success; factor B could be some measure of foraging efficiency where larger group size may lead to greater initial success, followed by a decline in fitness with increased numbers because of the need to share food among the group members. (*Modified slightly from Wilson, 1975.*)

a certain point, increased group size becomes a detriment to the fitness of each member despite the enhanced success of the group in competition with other groups.

Beyond this optimum group size, individuals are more fit if they either resist the recruitment of new members or if they leave themselves, thus effectively restoring the optimum. Therefore, as with so many other characteristics, natural selection apparently acts on individuals with regard to territoriality and group size, producing an optimum mix of benefits over costs. This general principle may well apply to all aspects of the sociobiology of group size, not just those related to territorial behavior (FIG. 9.3).

Another important generalization of sociobiology is that kinship considerations may figure significantly in virtually all aspects of social behavior, and group territoriality is no exception. Thus, all cases of group territorial defense known to me involve clans, troops, or direct extended families, in which the genetic relationships among the defending individuals are almost certainly greater than that between members of the contesting groups ($r > \bar{r}$). It should be emphasized that such relationships are not required by theory, but they are nice: Given the advantages of a certain modal group size, individual fitness could theoretically be raised just by aggregating, without regard to kinship considerations. But given the delicate balance being weighed between benefits and costs, the added "plus" of inclusive fitness appears to tip the scale in favor of kin. If you're going to benefit someone, you are more fit if that someone is a relative. These considerations are directly relevant to questions of human territoriality, because it is generally agreed that, territorial or not, our ancestors lived in groups within which coefficients of relationships were almost certainly high.

Note that ultimate explanations of the sort proposed here for territoriality in nonhuman animals neither exclude nor specify the proximate mechanisms that may be used. For example, size of territory can be adaptively decreased to maximize fitness when food is relatively abundant, by each individual simply spending more time obtaining food nearer its nest or den because it is now unnecessary to go farther, thereby permitting more individuals to establish territories in the same total space. In addition, more individuals may be attracted to regions of high food density, such that the heightened competition forces territories to be smaller. Such adjustments are opportunistic and adaptive, however they are achieved, and are found in a wide range of species.

To be sure, the sociobiological approach to territoriality or to any other behavioral phenotype does not ignore proximate mechanisms, although it usually does not concentrate upon them. They are simply different ways of approaching the same

PLATE 9.3 PRAIRIE DOG GIVING TERRITORIAL CALL NEAR THE ENTRANCE TO ITS BURROW. (PHOTO BY J. TAULMAN)

issue. To say that territorial behavior is adaptive is not to say that it is wholly instinctive in the sense that it must be entirely encoded within the genotype and resistant to experience. For example, young prairie dogs clearly *learn* the territorial boundaries of the family, the *coterie*, to which they belong; they are attacked when they wander across these boundaries. Once they begin uttering territorial calls, they are also attacked when they do so outside their territories but not when they are inside (King, 1955; Plate 9.3). There is also considerable evidence that prior experience within an area enhances the likelihood of successful defense of that area when the situation arises (Mackintosh, 1970). These accounts are independent of any discussion of the adaptive significance of territoriality and of each animal's susceptibility to learning such things in the first place.

We have already considered that the issue of whether humans are territorial, or dominance-structured, or innately aggressive, for that matter is largely one of definition, often laced with personal bias and a dose of social advocacy. A good case can be made in either direction. Our closest nonhuman primate relatives, chimpanzees and gorillas, engage in virtually no behaviors that are obviously correlated with territorial defense; as expected, competition within each of these species is low and resources are not economically defendable. On the other hand, the social carnivores such as lions, hyenas, and hunting dogs occupy ecological niches that are probably closer to those of our ancestors. And they do maintain group territories. Among present-day hunter-gatherers, some can be described as territorial, others are not. Adaptive considerations are equally equivocal when we consider whether territorial behavior would even be expected among *Homo sapiens*.

Unfortunately with regard to human behavior, it has for some reason become popular to equate *territorial* with *biological*. There can be no doubt that we are behaviorally the most flexible of all animals; how flexible, we do not know. It may be that a proprietary instinct lurks within each of us. However it may be that our biological nature leaves us with an almost totally open program where territorial behavior is concerned. In other words, we may have adopted a nonterritorial, chimpanzee-type strategy regarding the social use of space. If so, this may comfort the advocates of social learning theory and discomfit the die-hard hereditarians, but in fact territoriality is but one complex of strategies; there is nothing unbiological about not being territorial. Ask any chimpanzee.

What if You Succeed? What if You Fail?

As with competition for social dominance, there is no difficulty in predicting and understanding the consequences of success in territorial defense. The more interesting questions concern the losers. Territorial victors, whether solitary or in groups, can be expected to enjoy the fruits of their victory, the direct reproductive success and/or increments to inclusive fitness that confer adaptive value upon their behavior in the first place. But what of the losers? For their part, they should make the best of their bad situation. Of course, this involves different strategies for different species and, indeed, often different strategies for different individuals within the same species.

Perhaps the most basic strategy would be to forestall losing in the first place, an especially important consideration in cases where unsuccessful early attempts may prejudice the chances of later success. In such cases, selection may favor delayed sexual maturation, prolonged juvenile stages, or at least putting off entrance into the adult territorial arena until an individual is old enough to make success more likely. Furthermore, because males are more often the territorial sex, we can predict *sexual bimaturism* with males assuming adult behaviors later than females (Chapter 6; Selander, 1966; Wiley, 1974). Of course this strategy would be most appropriate for species that enjoy a long lifespan; waiting until next year to defend a territory is not a good evolutionary strategy if you are likely to be dead next year. Selection would also particularly favor delayed breeding attempts in proportion to the degree of polygamy of a species; if the sex ratio is 1:1, which it usually is, then the more unbalanced the adult breeding ratio, the more likely that a young member of the harem-maintaining sex will be excluded by older individuals. Furthermore, among highly polygamous species, it is more likely that the ultimate payoff will be high when and if that young animal becomes a successful territory defender itself. Note that these considerations apply as much to competition for social status as to competition for territories.

But what if you try to establish a territory and fail? One option is to leave the scene of one's defeat and try again somewhere else, hoping still to carve out a successful territory in a preferred habitat. Another is to remain as a helper and accrue inclusive fitness, as in certain K-selected species such as Mexican jays (Chapter 7). Yet another is to attempt to establish a territory in a suboptimal habitat or, finally, to skulk around as a nonterritorial *floater* near the territories of successful individuals. Presumably, this last strategy is made viable by the possibility of replacing a proprietor that weakens or dies. If territories are infinitely compressible, then the addition of more would-be territory holders to the same plot of ground simply results in each individual having a smaller territory. However, this would not be viable for the individuals concerned if territories that were too small failed to satisfy certain minimum resource requirements. At the other extreme, if territories were rigidly fixed in size, then a given area could support only a certain number of individuals, and any surplus animals would have to decide among the options listed above. In many cases, territories appear to be somewhat intermediate in their response to moderate crowding, resembling an elastic disk that can be compressed but only up to a point (Huxley, 1934).

One of the active and persistent debates in sociobiology concerns the possibility that territorial behavior serves to limit population density. Before considering this issue, a clarification is in order: It may be that a *consequence* of territoriality is that it restricts the number of breeding pairs and hence influences population size; this is different from territoriality having evolved as a proximate mechanism *in order to* limit population size in the ultimate sense. This latter interpretation requires either that territoriality has evolved for the good of the species or that it represents a curious combination of altruism on the part of nonbreeding, nonterritorial animals and/or spite (Chapter 4) on the part of territory holders. These latter individuals would be maintaining territories that are larger than they need, thereby reducing the fitness of others (the excluded contestants) as well as themselves by wasting time and effort in defending space that does not add to their fitness.

Because there is no compelling evidence supporting these latter propositions, we will simplify the issue and view territoriality as a behavioral strategy in which individuals engage as a means of maximizing personal fitness. As such, we may certainly inquire into whether territoriality influences population size, but we should bear in mind that any effects of this sort are probably indirect results of selection inducing each individual to maxi-

271

PLATE 9.4 ADULT MALE HERRING GULL PERFORMING THE "LONG CALL" AT HIS NEST SITE. This behavior serves as a long distance threat, a display that attracts mates and an indicator of possession of a mating territory. (PHOTO BY J. GALUSHA)

mize its fitness rather than a concerted strategy whose goal is population control for some larger and unspecified group.

Let us consider the predicted strategies for a seasonally territorial species in which helping is not a viable option. Theory suggests that the initial settlers should seek to establish territories in the best habitats, those conferring maximum reproductive success. This generally is what happens (e.g., Glas, 1960; Plate 9.4). Once this prime habitat is filled, unsuccessful contenders should begin to occupy suboptimal habitats (Brown, 1969). If this account is correct, total population sizes in suboptimal habitats should vary more than population sizes in prime habitats. For example, if 50 individuals can be accommodated in a prime habitat and another 50 in a secondary habitat, then a total population of 70 would distribute itself with 50 in the prime and 20 in the secondary, whereas a population of 95 would produce 50 (again) in the prime, because these desirable slots would be filled first, with the remaining 45 relegated to the secondary. There is support for this notion: Broad-leafed forest is prime habitat for European great tits, with pine forest definitely second best. The sizes of populations in the former vary less from year to year than they do in the latter (Kluyver and Tinbergen, 1953).

Finally, if all suitable habitats are occupied, any surplus must revert to nonbreeding, floater status (FIG. 9.4). There is considerable evidence for this as well, particularly among birds. Thus, some researchers attempted to evaluate the effects of songbirds in reducing populations of insect pests by shooting hundreds of birds from their territories in a Maine forest. They had hoped to compare bird-rich with bird-poor forest tracts with regard to their insect populations, but they were unable to do so because the newly vacant territories were immediately occupied by new owners (Hensley and Cope, 1951). This finding strongly suggests the existence of a large floating population of nonbreeders, ready to move into vacated territories whenever the occasion arises; see review by Watson and Moss (1970).

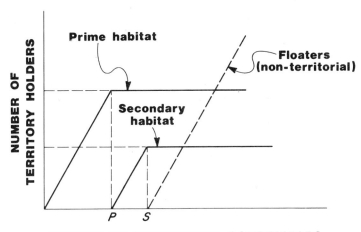

NUMBER OF TERRITORIAL COMPETITORS

FIGURE 9.4. A MODEL SHOWING THE FATE OF TERRITORIAL COMPETITORS. Initially all competitors seek to establish territories in the prime habitat, and all are successful. However, after the first P individuals have been successful the prime habitat is filled and secondary habitat is occupied. This habitat can accommodate only a limited number of individuals, so any additional would-be territory holders beyond a population size of S are relegated to nonterritorial, floater status.

273

One complication to this model is that habitat quality may not be a fixed parameter, independent of the animals occupying it. Rather, it may decrease, i.e., contribute less to fitness, with increased population density such that an individual territory holder may do as well in a suboptimal habitat, provided the density is low, as it would in an optimal habitat with a high density. Under this model, contestants for territory would choose the suboptimal habitat *before* the prime habitat is completely filled, i.e., less occupied, second-rate habitat may contribute more to fitness than one that is first-rate but crowded. At the balance point, individuals would be equally likely to reside in either habitat (Fretwell and Lucas, 1969).

In any case, there is abundant evidence that some individuals may be excluded from breeding by virtue of the territoriality of others. These floaters are most often males. For example, an eight-year study of red-winged blackbirds on a small marsh revealed that the number of breeding males remained more nearly constant than did the number of females (Davis and Peek, 1972), suggesting that extra males are excluded whereas additional females are accommodated within the territories of the successful males. If all females are breeding, then the population is reproducing at maximum capacity, regardless of whether some males are excluded; occasionally however, even females may be floating as well (Carrick, 1963).

The effects of all this on population size is difficult to predict. If we grant that environmental factors such as food and nest sites ultimately limit carrying capacity for any species and, further, that individuals will not maintain territories that are unnecessarily large, then even if females are excluded this may not be depressing the population size. In other words, if these excluded individuals breed, they might simply reduce the success of the given occupants, thus having no impact upon the total population size. Such a system would spread reproduction more equitably across all individuals, but there is no evidence that natural selection works that way; rather, each is out to do the best it can for its own inclusive fitness, and it is not surprising if some individuals lose out in the process.

"Losing out" may be immediate as well as long term. Individuals who are excluded from territory may suffer higher mortality (Watson and Moss, 1970). The situation varies from species to species, making further generalizations difficult. Territory size in European great tits is largely determined by the number of competing animals, so the effects of territoriality in limiting populations is weak at best. In turn, the number of competitors each spring is a function of the number that survived the previous winter (Krebs, 1971). Among Scottish red grouse, territorial behavior produces an excess population of nonbreeding floaters, with the number of floaters related to the average territory size, larger territories yielding more floaters. Territory size in turn is related to forage conditions of the preceding year (Watson and Jenkins, 1968; Watson and Miller, 1971).

These accounts of success and failure in territorial competition and its interaction with population levels has emphasized birds. Among mammals that lose out in territorial competition, dispersal may be a more common strategy than floating, perhaps because mammals are less well equipped than birds to monitor a number of territory holders for signs of weakness. However, excluded animals, especially males, are known for hyenas, lions, deer, seals, and many primates, to name just a few. They tend to congregate in bachelor groups and are often a source of instability for the establishment males. It is often not distinguishable whether these outcasts are the products of social or spatial competition. Of course it is only in the human mind that such distinctions are important anyhow; as far as the animals are concerned, each is simply doing the best it can. *C'est la vie.*

T E N

The Sociobiology of
Human Behavior:
Extrapolations
and Speculations

"Know thyself"
SOCRATES

More than two thousand years after the Socratic injunction, we are still strangers to ourselves. The scientific study of human behavior has traditionally been the province of social science: psychology, sociology, and anthropology, and, whereas great advances have certainly been made, there can be little doubt that *Homo sapiens* is not yet understood. Will sociobiology help? When Mark Twain was asked a difficult question: "I was gratified to be able to answer promptly, and I did" he wrote in *Life on the Mississippi*. "I said I didn't know." At this point, no one knows whether the perspective of evolutionary biology will shed light on our own social behavior, but it seems worth trying.

It may be argued that human behavior is so complex that biology will oversimplify and hence mislead. On the other hand, if human sociobiology is worth pursuing, it may be precisely *because* of that complexity. The task of unraveling such a difficult creature as *Homo sapiens* is so awesome that the social sciences need all the help they can get. They should leave no stone unturned, no tool unused in pursuit of anything that might offer further insight; and, considering the explanatory and predictive power of evolutionary theory when applied to the social behavior of other living things, application of this evolutionary approach to human behavior seems eminently sensi-

276

ble. If biology seems arrogant in seeking to include humans within its scope, think of the greater arrogance of a social science that refuses help when it is offered.

Actually, biological approaches to human behavior are not altogether novel. For example, in recent years there has been a spurt of interest in human ethology. This approach has been characterized by direct observation of people behaving in natural situations, free of any experimental contrivance (Blurton-Jones, 1972; McGrew, 1972). Beyond this, researchers have begun applying ethology's conceptual baggage of *releasers, fixed action patterns*, and *innate releasing mechanisms* to the study of human behavior (Eibl-Eibesfeldt, 1975). Efforts of this sort may well prove valuable, but it is at least possible that even greater insights will ultimately be achieved by combining the observational approach of ethology with the predictive tools of modern sociobiology. Thus, a worthwhile endeavor for students of human sociobiology would be the making and testing of predictions based on the assumption that individual humans beings behave in such a manner as to maximize their inclusive fitness.

Because this territory is almost entirely unexplored, the approach in this chapter will be frankly speculative, suggesting some direct applications of the principles of sociobiology to a new species, ourselves. The preceding nine chapters have presented empirical observations concerning the sociobiology of nonhuman animals as well as the fundamentals of theory. Such theory has not been developed with any particular species in mind, and it may well be generalizable to all living things. This chapter will be especially concerned with exploring possible implications for human social behavior, *assuming* that our species follows these same general rules. This caveat is important. We are going to play "Let's Pretend", and see where it takes us. In other words, if the Central Theorem of Sociobiology holds for humans, and we tend to behave so as to maximize personal, inclusive fitness, then what? We can begin by making predictions that may ultimately be confirmed or refuted. At least, it is a start.

In nearly all cases these will be real predictions; their validity is simply unknown. The future should provide exciting oppor-

tunities for evaluating these and other extensions of sociobiological thought.

Some Problems and Justifications

Humans are unique among animals, both in terms of their
characteristics as objects for scientific scrutiny and also in terms
of the social consequences of such efforts. Objections have been
raised to the application of sociobiology to human behavior,
much of it directed against Harvard zoologist, Edward O. Wilson, whose book, *Sociobiology: The New Synthesis*, crystallized
many of the issues and served as a rallying point for both proponents and opponents; see Allen et al. (1976) for a particularly
vigorous attack. Criticism appears to be focused on two major
themes: The social and political implications of human
sociobiology and the question of scientific validity for such research. These are important and emotion-packed issues, and
they warrant attention before proceeding with the meat of this
chapter. Like sociobiology itself, the controversy involves complex issues that can only be touched upon here.

Concern has been expressed that human sociobiology represents racism in disguise: This is simply not true. Sociobiology
deals with biological universals that may underlie human social
behavior, universals that are presumed to hold cross-culturally
and therefore cross-racially as well. What better *antidote* for racism than such emphasis on the behavioral commonality of our
single species?

Sociobiology searches for the biological foundations of social
behavior. When evaluations of this sort are made for human
behavior and supporting evidence is presented, this may be
misread as somehow *condoning* the behaviors in question.
Again, this is nonsense: Ethical judgments have no place in the
study of human sociobiology or in any other science for that
matter. What is biological is not necessarily good, assuming
here that human social behavior is in fact found to have a valid
evolutionary substrate. Diseases are part of our biology; this
does not imply that they are good. We study pneumococci, seek-

ing further understanding of how they are put together and why they do what they do; this does not imply that we approve of pneumonia. Some critics have even gone so far as to advocate the suppression of research in human sociobiology. Restraints upon research and attempted prohibitions against inquiry smack of totalitarianism and book burning; they should be anathema to a free society. Human sociobiology will hopefully proceed and with adherence to scientific validity alone. There must be no forbidden knowledge. "Knowledge humanely acquired and widely shared, related to human needs but kept free of political censorship, is the real science for the people" (Wilson, 1976). Science is neither good nor bad of itself; only the socially mediated disposition of such knowledge is susceptible to moral judgment. And yet science can be misused.

Unfortunately there are no easy solutions to this dilemma. The conceptual tools of sociobiology have been employed successfully on nonhuman animals, and this can only whet the appetites of those of us who recognize how little we understand our own species. Most biologists are committed to the proposition that a herring gull is well worth knowing in its own right. But so is *Homo sapiens*. We are unlikely to deny ourselves the same analytic attention we lavish on other species, and there is no reason why we should.

Another criticism of human sociobiology that is essentially ethical or, rather, political in nature is that it tends to portray social systems as natural and therefore provides support for their continuance. Critics have furthermore pointed out that the systems being supported tend to be those within which sociobiologists, as private citizens, are functioning successfully—capitalist, sexist, etc. First, we have already emphasized that *biological* and *good* are not synonymous. Furthermore, concern with adaptive significance need not imply a Panglossian view that "all is for the best in this best of all possible worlds". Population geneticists have employed the concept of *adaptive landscape* in which populations are seen as isolated specks on a complex, multidimensional landscape of numerous peaks of different shape and height, separated by valleys. The peaks represent adaptation, and so the greater the altitude of any population, the greater its average fitness. Populations move upwards to greater fitness by natural selection operating upon each component individual; they may also fall over a cliff, i.e., become extinct. But these movements are essentially blind. Therefore an individual or a species has no idea where it is relative to the rest of the landscape and accordingly cannot tell if it would be better off (more fit) on a different peak entirely. Because its predominant movement is upwards, it may be restricted indefinitely to a small hillock, unable to attain a much larger, more adaptive, mountain nearby from which it is separated by an uncrossable valley. For this reason, it has been suggested that maximum adaptation may require the addition of occasional nondirected, random movements, i.e., genetic drift, in order for populations to escape from their adaptive prisons to the greater heights of enhanced fitness (Wright, 1969).

For those who see politics in every aspect of the natural world, perhaps the preceding is an argument for anarchy, at least in limited doses. But, on a more sober level, this model shows how evolutionary biology does *not* claim that all phenotypes must represent fitnesses that are absolutely maximal. Thus, sociobiologists are not necessarily apologists for the status quo; the whole issue is a red herring. In fact, historians of science might well find this debate amusing. When evolution by natural selection was originally proposed, it met strong resistance from the entrenched European political and social establishment. Evolution implies movement and susceptibility to change. It sees the living world in a constant state of flux, as opposed to the static system of divine creation. Coming on the heels of the American and French Revolutions, such a radical world view was distinctly uncomfortable for the powers-that-be. Now, purveyors of evolutionary thought are accused of defending social intransigence. How strange.

In the hope of avoiding further confusion, an important distinction should also be made at this point: the difference between *analogy* and *homology*. Structures are analogous if they serve similar functions, often because of similar selective pressures. The wings of a bird and of an insect are classic examples: Both are flapped in the air and both contribute to flying. But the similarity is superficial only. An insect's wing derives from an out-folding of the body wall, whereas a bird's wing is derived from the same bone structures that give rise to a bat's wing or a man's arm. These latter structures are all homologous: They possess an underlying similarity based on common evolutionary descent. Like the wings of birds and insects, similar behaviors of human and nonhuman animals may represent analogy and nothing more, or, like the wings of birds and bats, they may be homologous as well. Basically, our ignorance on this important matter is due to the fact that human behavior derives from many possible sources. If a particular pattern resembles that found among nonhuman animals, it may occur among humans almost entirely as a function of learning and social tradition, in which case it is analogous only. To be homologous, it would have to derive from the action of similar genes.

This poses a serious problem for the study of human sociobiology, because we cannot experimentally manipulate genes and vary environments at will among humans so as to isolate causative factors, as we can among nonhumans. Therefore we must content ourselves with descriptions of what actually occurs in human behavior, often being unable to parcel these phenotypes to social experience or evolution. For example, among certain Eskimo societies, it is customary for the very old to sacrifice themselves in times of famine. This is in accord with sociobiologic predictions, particularly the biology of altruism. Thus, kin selection could evolve this tendency when it increased the reproductive success of surviving relatives. Furthermore we would expect that old people especially would be selected for such behavior, because their ratio of benefit to cost is high; they are likely to have many relatives who benefit. Simultaneously, because of their senescence they have little potential reproductive future; i.e., low cost associated with their act. On the other hand, such behavior may also result from cultural tradition, spread among Eskimo families because it is adaptive, that is, families with such traditions are more likely to succeed and to generate additional families with similar traditions than others whose cultural practice precludes geriatric self-sacrifice.

Accordingly, culture can mimic biology and vice versa; and it may be difficult or even impossible to establish the precise relevance of natural selection in such cases. However we can use our understanding of the selective process to predict adaptive behaviors; our strongest evidence for the legitimacy of human sociobiology may well be the documentation of cross-cultural behavioral universals. Thus, because there is no way to eliminate culture as a confounding variable, the next best thing may be to examine human behavior in a wide array of societies. This would essentially be an experiment in which our biological nature as the species *Homo sapiens* was held constant, while cultural practices were permitted to vary.

The persistence of common patterns, despite a diversity of cultural overlays, may reflect underlying biological tendencies consistent with but not proving the action of natural selection. Such evidence would be particularly cogent if these universals conformed to the predictions of evolutionary theory; ideally, these predictions should be generated without any advance knowledge of what is really true. In each case, the particular pattern may be interpretable in proximate terms as the outcome of particular cultural factors, but consistent findings could also at least suggest an ultimate, biological reason for *why* these cultural factors exist and why *Homo sapiens* responds to them as they do, rather than in some different way.

The Eskimo example may provide an opportunity for the evolution of altruistic behavior through genetic assimilation in conjunction with kin selection. Assume that geriatric self-sacrifice is initially a phenomenon of cultural tradition and nothing else. Selection could then operate upon the *susceptibility* to such teachings, assuming here, that susceptibility genes of this sort exist. Eventually, the behavior could be genetically incorporated and thereafter maintained by kin selection. When biology and culture are confounded, we cannot yet identify the root cause, but we can and should continue to speculate and gather evidence, so long as we do both with open minds.

The Hebraic prohibition against eating pork is a clear-cut

example of an adaptive, culturally mediated behavior; it undoubtedly reduces the incidence of trichinosis, and it undoubtedly is not genetic in origin. But there is at least the possibility that even dietary preferences have some underlying genetic basis. Among human populations such as Orientals and Australian aborigines who consume few or no dairy products, the enzyme lactase, which acts to digest the sugar lactose found in milk, tends to be missing (Kretchmer, 1973). As a result, the milk sugar, lactose, cannot be digested, and ingestion of milk commonly causes intestinal distress. Cultural practices may have selected for the presence of lactase among herding peoples; or the fortuitous presence or absence of the relevant genes may have predisposed different societies to different cultural norms. Certainly just because something is adaptive does not mean that it must be biological even if it accords with sociobiologic prediction. But it is equally true that being cultural does not require that it cannot be biological *as well*.

Sociobiology relies heavily upon the biology of male-female differences and upon the adaptive behavioral differences that have evolved accordingly. Ironically, mother nature appears to be a sexist,* at least where nonhuman animals are concerned. There may or may not be similar biological underpinnings of sexism in human societies; we do not know. Certainly, analysis and evaluation of male-female differences in human behavior should not be construed as supporting its propriety. Human sociobiology seeks to explore our nature, not to legitimize our foibles. Once again, it seems scientifically appropriate and ethically responsible to employ theory to make testable predictions. If reality accords with theory, this does not prove anything, because social and cultural factors may mimic the action of natural selection regarding sex differences in behavior just as with Eskimo altruism. Human social behavior is the product of many interacting factors and it is certainly unlikely that sexism is entirely biological. But it also may not be entirely cultural. One way of shedding more light is to make some predictions and try to interpret the realities.

---

*Sociobiology is sexist if sexism is recognition of male-female *differences*; however, it does not imply that either sex is *better*.

A final, political concern of sociobiology's opponents is that it represents an inimical form of biological *determinism*. This worry may derive largely from confusion of genetic determinism with genetic *tendencies,* and it also leads into the primary scientific complaint against human sociobiology.

Genetic Predispositions and Human Behavior, or Why Is Sugar Sweet?

Evolution operates by changes in gene frequencies over time, and natural selection is the means by which adaptation is achieved. Because evolution can operate upon a phenotype, including behavior, only if some correlation exists between that phenotype and the genotype, any consideration of adaptive aspects of human behavior must assume that genetics plays at least some role in mediating human behavior. But this is not to claim that we are involuntary automatons, prisoners of our DNA. Thus, we may be predisposed to behave in a certain manner without being at the mercy of heredity.

Male stickleback fish have a red breast in the spring when courtship and mating occur. Such males respond with automatic aggression toward other males, so long as they also possess a red breast. Indeed, the intruder need not even be recognizable to us as a fish at all; almost any red object will produce this genetically coded response, and this will occur even if the animal has been kept in total isolation and therefore has had no opportunity to learn the behavior. Such rigidly stereotyped, mouse-trap-type responses are common among nonhuman animals, especially invertebrates, fish, reptiles, and birds, but not in humans. We do not experience anything like this genetically mediated automaticity of response. We may feel vague and general inclinations to respond in certain ways, but such tendencies can be overridden consciously and modified drastically by our experience.

If quiet, romantic little coffee houses are not regularly the scene of passionate lovemaking under the tables, it is generally not because the biological inclinations aren't there; the patrons simply are also responding to cultural taboos that at least in this

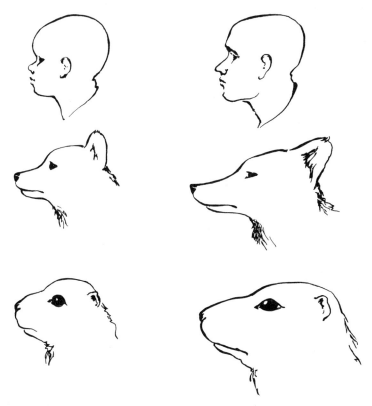

FIGURE 10.1 GENERALIZED SENSITIVITY TO BABYISH FEATURES. Humans typically feel more inclined to cuddle and care for individuals who possess certain features such as blunt stubby body parts and relatively large eyes and head compared to the rest of the body. (ILLUSTRATION BY D. COOK)

situation are stronger. But the fact that such inclinations can be suppressed does not mean that they do not exist. For example, humans have almost certainly been under intense selective pressure for responsiveness toward their infants. Although we do not possess the automatic inborn response patterns used by so many nonhuman animals in recognizing their young, we may well be responsive to certain general physical patterns that characterize infants: Large eyes, rounded features, unsteadiness of gait, etc. (FIG. 10.1). We can force ourselves to remain unresponsive to such patterns, but the very fact that force is required suggests a natural inclination in the other direction.

In suggesting possible adaptive predispositions in human behavior, this is the type of system sociobiology has in mind: A flexible, modifiable and perhaps rather fragile set of inclinations but a potentially significant part of our biology, quite possibly grounded in the evolutionary wisdom of our past.

This approach is clearly at variance with many other views of human nature. Simone de Beauvoir described human beings as *l'être dont l'être est de n'être pas* [the being whose essence is having no essence]. Behaviorism is one of the dominant schools of American psychology; part of *its* essence is captured in the following, oft-quoted challenge from J. B. Watson (1930), its prime mover.

> Give me a dozen healthy infants, well-formed, and my own specified world to bring them up in and I'll guarantee to take any one at random and train him to become any type of specialist I might select—doctor, lawyer, artist, merchant-chief and yes, even beggarman and thief, regardless of his talents, penchants, tendencies, abilities, vocations, and race of his ancestors.

Choice of a profession may well be more specific than the genetic predispositions that sociobiologists would claim. In an evolutionary perspective, the only concern of all human behavior is the production of other humans. Behaviors evolve when they confer greater fitness than their alternatives, although in many cases the route from behavior to fitness may be tortuous and the adaptive significance obscure. However this is not to say that our behavior is necessarily directed by our genes. As explained earlier, DNA is only a blueprint, a potential for constructing something whose actual appearance may vary dramatically as a function of the materials available to do the job. Not only that; even after a behavioral structure is completed, it may assume a dramatically different shape from one situation to the next. Nevertheless some correlation always exists between genes and behavior, even human behavior. It may be precise, as in the nervous system wiring that produces the blink reflex in response to sudden loud noises. Or it may be diffuse and therefore almost entirely dependent upon environmental influences, as in the case of personality. It may be nothing more that a restriction of potential: A human being cannot develop the personality of a seahorse, just as a seahorse cannot be loving, happy, greedy, jealous, or schizophrenic in the same sense that we can.

Practitioners of human sociobiology therefore do *not* necessarily advocate biological determinism of human behavior. The difference between determinism and genetic *influence* is the difference between shooting a bullet at a target and throwing a paper airplane; the paper airplane is acutely sensitive to environmental influences such as wind, and its ultimate path is not entirely predictable by the thrower. The great evolutionary biologist, Julian Huxley, warned against "nothing butism," the idea that because humans are animals, it necessarily follows that we are nothing but animals. This is important advice, but it also applies in the other direction: We should beware the notion that humans are nothing but the outcome of their experiences. It stretches credulity to imagine that *Homo sapiens* is unique among animals in possessing no biological components relevant to social behavior. If nothing else, we must possess a genetically mediated *capacity* for culture, learning, language, etc. in the same sense that wolves have the capacity of developing complicated dominance relationships. Furthermore, we lack the capacity to perform accurate long-distance migration, found in many birds, just as herring gulls are incapable of forming political parties or writing a symphony. The next step, at which many sociologists, psychologists, and anthropologists will doubtless balk, is to identify predispositions for the learning of particular things, the acquisition of cultural traits that, despite their diversity, demonstrate a certain functional, adaptive, consistency. In some cases, learning may be minimal or even largely unnecessary; in others, it may be overwhelmingly important. In any event, sociobiology enables us to postulate an end toward which these diverse means might be directed: Much of human behavior may be adaptive; i.e., it may maximize the inclusive Darwinian fitness of each individual.

Unfortunately, most of the thought concerning genetic aspects of human behavior has been highly emotional and hence, extremist in each direction. On the one hand, many social scientists advocate that humans are born as "tabula rasa", blank slates with infinite potential to become anything at all, depending only on experience. On the other end of the spectrum is extreme genetic determinism, in which free will and the capacity for self-betterment is denied, and we are literally prisoners of our heredity. As usual in such outlandish debates, the truth doubtless lies somewhere in between (Dobzhansky, 1976). For our purposes, this means that we are justified in assuming an *influence* of genetics in human behavior, although not necessarily a controlling influence. Sociobiology's approach is to ask natural selection about the expected direction of that influence.

Social scientists are likely to be uncomfortable with this approach, based as it is upon the notion of genetically influenced, although not genetically determined, behavioral predispositions. However such thinking is not really novel for social science: The role of genotype is basic to the theories of the great developmental psychologist Piaget (1973), with his emphasis upon the adaptive unfolding of strategies for dealing with experience, and Chomsky's conception of an underlying "innate deep structure" in human language (1972). Both of these approaches recognize a biological substrate for behavior. But it is one thing to identify certain underlying givens in the human repertoire; it is quite another to use evolutionary theory to explain the reason for these givens and then to go even further and predict the existence of other tendencies, perhaps some not yet identified. This is the unique promise of sociobiology.

In many cases, genetic influences upon our behavior are so subtle that we take the results for granted—it sometimes takes considerable mental effort to step back from our behavior and view ourselves objectively. For example, why is sugar sweet? Of course, chemists have identified similarities in the molecular structure of those compounds we call sugars. But sweetness does not reside in these molecules; rather, it is in our taste buds, our response to these substances. To an anteater, sugar is not desirable eating, but of course ants are a different matter. To us, ants taste bitter. Why the difference? Our primate ancestors probably ate a great deal of fruit, which is more nutritious when it is ripe and rich with sugars. We were therefore selected for responding positively to the taste of those chemicals that characterize ripe fruit. Accordingly sugar tastes sweet to us; an articulate giant panda would probably say the same about bamboo shoots.

In a sense humans are unique among animals in that our behaviors appear liberated from the tyrannical demands of biology, mediated by natural selection. We can substitute personal satisfactions for adaptive significance. But carry this thought one step farther: What behaviors do we find satisfying, and why? Sex, good food, rest, the respect of others, physical comfort, personal power and autonomy, coordinated and successful movements (athletics, dancing), the accomplishments of ourselves and our offspring, all these pleasures contribute eventually to our own fitness, and therefore we have been selected to engage in them. We find them sweet.

Armed with the intellectual paraphernalia of sociobiology, and cautioned as to the uncertain role of genes in human behavior, let us now attempt a preliminary, speculative and predictive excursion into human sociobiology.

Mate Selection

Assuming that human behavior has been selected to maximize inclusive fitness, we would expect to see evidence of this in choice of a mate. Sexual attractiveness of a potential partner is not different from sweetness of a potential food: The quality is not inherent in the subject perceived but, rather, in the perceiver. Just as with our choice of foods, we should choose mates that are good for us. Therefore, insofar as we have been selected for positive responses to certain physical characteristics of potential mates, selection would favor those individuals preferring mates who conferred maximum reproductive success upon the discriminating partner. The situation is analogous (homologous?) to sexual selection in other animals (Chapter 6): Individuals are most fit if they make the best choice. Therefore, we could predict a positive correlation between physical viability, general health, and other indicators of reproductive potential among members of each sex and those characteristics perceived as attractive by members of the opposite sex. Regularity of features, smoothness of complexion, optimum stature (neither too short not too tall), good physique, all of these things count, and for good reason. Among the secondary sexual characteristics of women, breasts help nourish offspring, although within normal limits breast size is not correlated with milk production, and hips of a certain minimum width facilitate childbirth. In this regard the traditional Mother Earth Goddess of antiquity may come closer to biological appropriateness than does today's chic svelte look. If in earlier times, plumpness implied command of resources, in American society at least, it no longer does. Even this change in cultural norm is at least not maladaptive because modern medicine has relaxed the biological requisites for reproductive success: we have hospitals, cesarians, and baby bottles.

All human societies have prohibitions against incest between siblings and between parents and offspring. Social scientists commonly interpret these universals as due to the potential for family disruption inherent in sexual rivalry between close relatives (Levi-Strauss, 1969). Of course close breeding would also reduce the fitness of the participants because of the reduced viability of the offspring produced. Like the Jewish dietary laws, incest avoidance is cultural. It is also adaptive. However, it goes further in being universal and therefore suggesting genetic predispositions. The anthropological literature on incest avoidance systems is immense; for a biological perspective, see Fox (1972).

Humans are mammals. We therefore experience the same differences between male and female parental investment as most other animals (see Chapter 6). Extrapolating from the theory, men should accordingly be sexual aggressors. Men should also compete among themselves for the sexual favors of women, and women should in turn be selected for assessing men in large part by the quality of the reproductively relevant resources they control. It may be significant that women tend to concentrate their reproductive interests upon men who are socially, economically, and educationally above them (Blake, 1971).

Savior-faire and the capacity to command respect from others are characteristics of men that have long been valued by women. The reverse is much less prevalent and, in fact, too much competence and accomplishment by a woman is often threatening to a prospective male partner unless his ego is especially secure. Successful single women often face a problem in that their range of prospective mates tends to be limited to those above them in the socioeconomic scale. This adaptive tendency for women to marry up (hypergamy) would occur even when reproduction is not intended or even if it is consciously excluded. Thus, such basic adaptive inclinations as hypergamy may be insensitive to personal conditions (i.e., intentional childlessness), which are themselves too recent to have been acted upon by natural selection.

Similar tendencies among animals often result in sexual bimaturism (Chapter 6) and indeed, it is almost universal in human societies that men are older than their mates; furthermore, puberty occurs earlier in girls than in boys. Older men are more likely to control resources of value to a reproducing female and, given this correlation, selection could easily result in female preference for a degree of male maturity. There is nothing very peculiar in a 70-year-old man marrying a 25-year-old woman. Even if tongues wag, there may be a grudging unconscious recognition that both could be maximizing their fitness. A young man with an old woman seems much more inappropriate.

In fact, similarly adaptive human predispositions could be responsible for the prolongation of reproductive potential in men relative to women. Female fertility ends at menopause, roughly at 45 to 55 years of age. In contrast, men can father children at any age. The physical strains of childbearing may have selected against reproduction by older women, especially if they are able to enhance their inclusive fitness by contributing indirectly to

291

the eventual reproductive success of existing children, grand-children, and other relatives. Increased fragility with increasing age thus adds to the cost of reproducing, while at the same time an accumulation of wisdom and experience adds to the ultimate reproductive benefit that grandmothers, great-aunts, etc., can provide, even if they now assume a nonreproductive role themselves. In contrast, the physical cost of sex is minimal to men; they do not get pregnant. This, combined with the general correlation of age with control of resources and the pre-sumed female preferences with which this correlation is asso-ciated, could select for the relatively prolonged sexual potency of men.

The polygyny threshold model of female mate selection (Chapter 6) suggests that maximization of female fitness re-quires discriminative mate selection. Nonhumans apparently achieve this by weighing the disadvantages of sharing a male's attention against the advantage of good resource quality often provided by an already-mated male. Among polygynous human societies the number of wives allotted to a man is a func-tion of resources; i.e., he cannot have more wives than he can adequately support. This phenomenon appears to be entirely cultural, but it is also adaptive, and the parallels with other animals are striking. In addition, younger men in human polygynous societies are almost invariably monogamists, if they are mated at all. Polygyny is reserved for the older men who control greater resources.

Furthermore, much of human courtship can be interpreted as providing opportunities for mutual assessment, analogous (homologous?) to one of the suggested functions of courtship among nonhuman animals. Expenditure of money on dinners, flowers, movies, or gifts provides a means of displaying finan-cial resources and willingness to dispense them. Dancing and athletics demonstrate physical prowess, along with the display of trophies or other symbols of successful rites of passage. Living-together arrangements permit assessment of sexual compatibility.

I have pointed out (Chapter 6) that the biology of male-female differences predisposes males of most species to sexual aggressiveness, advertisement, and availability, while females are selected for discrimination and sales resistance. Humans are perfectly good animals in terms of their reproductive biology and the necessarily high parental investment required of women compared with that of men. Accordingly our species may also demonstrate the same adaptive difference in male-female reproductive strategies, albeit modified by such cultural phenomena as laws, moral norms, and experiences during childhood. Prostitutes are nearly always female in every culture; pornographic books and movies are directed almost exclusively at men; men are readily seduced by explicit sexual advances whereas women are more likely to be influenced by a romantic setting, some indication of genuine commitment by the man, etc.

A man's sperm represents very little investment; a woman's egg, while small itself, may have enormous significance to her. Given the peculiar biology of *Homo sapiens*, males probably maximize their fitness by a degree of reproductive, parental, commitment to their mate(s). However, optimum male strategy would include remaining susceptible to additional copulations, so long as they did not require further investment. Women may also have been selected for an interest in copulations outside the pair-bond but, because of their greater involvement in the consequences of such activity, women should be more fussy than men. Men are predicted to feel more threatened by the sexual activities of their women than women should feel as a result of sexual dalliance by their men. In terms of Darwinian fitness, men *are* threatened more than women by out-of-pair-bond copulations by their partner. In effect, I am suggesting a potential biological basis for the double standard. Once again, I am *not* suggesting that what is biological must also be ethically right, or unavoidable. With the current changes in social tradition and the perfection of birth control, we can certainly expect dramatic changes in such traditional behavior patterns. It will be interesting to see what evolves from modern society. We can afford equality of sexual behavior only when the biological consequences of such behavior are equal, but so long as women, not men, get pregnant, some differences can be predicted.

Responses to Behavior of the Mate

Few decisions are irrevocable. The behavior of mates toward each other can be expected to vary with many factors, especially as situations change and as life experiences provide further information regarding the partner. Relevant information could include actual changes in physical condition, behavioral predispositions, etc., or more accurate assessment of existing factors, if either partner had previously received an inaccurate impression of the other. These inaccuracies could be due to deceit, insufficient perceptiveness, or both. In any case, sociobiology suggests that one essential criterion could underlie the patterning of responses to the behavior of one's mate: The consequences of such responses or the failure to respond for the fitness of the individual concerned. Thus, mate A should be particularly sensitive to any characteristic of mate B, either physical or behavioral, that has important consequences for the inclusive fitness of mate A. Furthermore, the response of mate A should be such as to enhance or maintain his (or her) personal fitness and not that of B.

Sociobiology accordingly predicts, for example, that divorce or its equivalent correlates closely with any substantial decrement to successful reproduction. Failure to consummate a marriage is often sufficient cause for annulment, and impotence and infertility have historically been a cause of marital breakdown. Among kittiwake gulls, over 60% of the pairs retain their mates of the previous season (Coulson, 1966). These old married couples begin reproduction several days earlier than "newly weds". They appear to benefit from the increased coordination that comes with mutual familiarity. But "divorce" also occurs: Pairs that failed to hatch any young the previous year are three times as likely to change partners as are those that bred successfully. This represents an adaptive strategy: Find a new mate if you and your old mate are reproductively incompatible. It would be interesting to learn whether divorces among humans are more likely following the death of a young child, for example.

Let us assume that the major biological function of man-woman pair-bonding is the production of successful offspring. Love, compaionship, and sexual satisfaction can all be seen as proximate means of achieving this ultimate end. We seek them and find them pleasurable for the same reason we find sugar sweet. This certainly explains the prevalence of couples remaining together for the sake of the children, even when the adults are personally dissatisfied. Of course, such adult self-sacrifice is ultimately selfish, in that offspring are the primary vessels carrying the adult genotype, their most direct route to Darwinian fitness and evolutionary success.

Significantly, decreased reluctance to divorce in present day America correlates with increased availability of day-care centers and other forms of substitute parenting. It is axiomatic that divorce is easier when there are no children. Admittedly, this commonplace observation may reflect true altruism by the parents or the selfish recognition of the increased burden of single parenting. But it may also represent our own deep-seated awareness that such an action often runs counter to the biologically appropriate reproductive strategy of an animal whose offspring have a long period of dependency. For the same reason that sugar tastes sweet, we can expect divorce to taste bitter, especially if there are children.

What about the sociobiology of cuckoldry? The male response to apparent female adultery in mountain bluebirds (Chapter 3) reflects an adaptive male response, aggressive intolerance. This strategy is biologically appropriate, as long as replacement females are available and the chances of rearing a brood are high and/or if the threat of such response causes the potential adulterer to have second thoughts. No information is currently available regarding the *female* response to *male* adultery among animals, although I predict it to be significantly less intense. This is because females of every species enjoy an enormous biological advantage: They know that they will share 50% of their genes with each of their offspring. Males lack this assurance. Accordingly, the philandering husband is no great disgrace, as long as he also provides adequately for his domestic obligations. His fitness may be high. The cuckolded husband, on the other hand, is an object of ridicule. His fitness is probably low.

Sociobiology predicts that human males are therefore significantly more intolerant of infidelity by their wives than wives are of their husbands. Violent crimes are often precipitated by domestic infidelity and it is no secret and should be no surprise that adultery is punished with particular severity in male-dominated societies. Significantly, socially approved wife sharing among unrelated men occurs only in certain societies, such as some Eskimos, in which the rigor of the natural environment makes cooperation of greater value than absolute confidence of genetic relatedness. Finally, the marital system of commitment and fidelity may provide a means of mutual restraint, with each partner threatening potential infidelity in the event of such behavior by the other.

On one level, this discussion of human sexuality in terms of strategies for the maximization of fitness may appear to be missing the boat. Sex is great fun. That's why we do it, isn't it? In fact, our pleasure in sex is but another example of the "why sugar is sweet" argument; the enormous satisfaction of sex may be viewed as a trick by natural selection to ensure that we reproduce. But beyond this we humans are unique among animals in the additional use we make of sex. Sex is an expensive endeavor, and it is energetically demanding and exposes the participants to increased likelihood of predation. On the other hand, it is adaptive in serving reproduction, and clearly that is why it is done. And among nonhuman animals, that is virtually the only reason. They copulate almost exclusively with regard to reproduction; sex is a serious, businesslike activity, performed when there is a high probability of offspring being produced. On the other hand, humans appear to be unique in that we engage in *nonreproductive* sex; unlike other animals, human sexual activity is not limited to distinct breeding seasons or special periods of receptivity (estrus or heat in other mammals) correlating with ovulation by the female. Human sexuality has essentially been liberated from its domination by hormones, and accordingly we engage in sex throughout the calender, if not around the clock.

But for what has it been liberated? One guess is that our novel use of sex relates to the unique problem posed by our utterly dependent infants. Human infants are totally helpless and require the committed attention of one parent, invariably the woman, since she is also adapted to nourish her newborn. It would certainly help if there was a daddy around to hunt, scavenge, defend the female and her child, etc. Given that, during our evolutionary development, offspring were more likely to be successful if they received the committed assistance of at least two adults, selection would favor any mechanism that kept the adults together. Sex may be such a device, selected to be pleasurable for its own sake, in addition to its procreative function. This would help explain why the female orgasm seems to be unique to humans; among other animals, reproduction is the only goal, and satisfaction per se is irrelevant. In addition, loss of estrus among humans contributes to sexual consistency that may in turn help maintain a stable pair-bond. This interpretation, while admittedly speculative, is clearly at odds with the traditional Vatican argument that nonreproductive sex is somehow animalistic and dehumanizing. Sex for its own sake may in fact be one of the few biological distinctions of our species, a uniquely human attribute.

Strategies of Being a Parent

I have already discussed a study of parental defense in Alpine accentors (Chapter 7). These small birds show increased bravery in defense of their offspring as the latter grow older, an adaptive strategy because with increased age the chances of further reproduction decline, rendering the offspring at hand increasingly important to the fitness of the parent(s). Similarly, white-crowned sparrows of the *pugetensis* subspecies that are capable of renesting show less parental defense than does the *gambelli* subspecies with only one brood per season and no second chance (Chapter 7).

Considerations of this sort yield several predictions concerning human parental behavior. Thus, abortion is generally viewed with increasing repugnance as the embryo becomes older. We have relatively few qualms about destroying a just-fertilized egg but feel differently about a seven-month fetus. The older the offspring and the greater the investment, the greater the defense. Killing or abandoning a newborn baby is usually even more difficult. After all, the mother has felt the child grow and develop, and there are powerful proximate mechanisms operating to ensure continuation of parental care once so much has been invested. Extend the analysis: If parents were forced to choose between their newborn baby and their three year old, the three year old would almost certainly win. The older child has successfully completed a hazardous time, infant mortality being relatively high, and it therefore represents a better bet for evolutionary success than does the newborn. A decision in its favor would therefore be adaptive for the parents. Certainly in most cultures, little fuss is occasioned by the death of an infant, as opposed to adult mortality.

Of course, one might argue that this choice has nothing to do with adaptive strategies and the maximization of Darwinian fitness; we choose the three year old because we have grown to love it, whereas we hardly know the infant. But that is the point: Why do we *grow* to love a child? Maybe because the older the child, the more value it represents in terms of our evolutionary future. Like love between adults, parental love is highly adaptive. In biological terms, parents who decide to invest preferentially in such offspring will ultimately produce more successful offspring than will parents who follow an alternative strategy; i.e., the former will be more fit and such tendencies could therefore be selected. In this case, we identify as *love* the proximate mechanism ensuring such tendencies. Human love, then, could be a universal behavioral means to a biological end.

Let us assume that human parents demonstrate solicitude toward their children according to the basic formulations of kin selection (Chapter 4): Given that genetic relatedness between the parent and each child is equal, solicitude toward each child should vary inversely with the risk to the parents' reproductive success via other born or unborn offspring and relatives. In addition, parental solicitude should vary directly with the extent to which the assisted child will benefit reproductively by each investment. Thus, a deformed or defective child should generally receive little investment or even suffer infanticide. This is the predominant pattern throughout most human societies. An older child who has survived the dangers of infant mortality should warrant greater investment than one less likely to provide an evolutionary return. Similarly, we should reserve our greatest antagonisms toward unrelated individuals, especially when they have achieved maximum reproductive value, at adolescence (Emlen, 1966).

Parental investment has been defined as any behavior that increases the chances of survival and reproduction by offspring at the cost of the parents' ability to invest in other offspring (Chapter 6). Parents with less future reproductive potential should therefore be selected for greater parental investment in current offspring. There is good evidence for such a tendency among animals; witness the *gambelli* versus *pugetensis* white-crowned sparrows, and a similar pattern can be predicted among humans. For example, compare the decisions made by two women, pregnant for the first time, and both hypothetically faced with the choice of saving themselves or their baby. One woman is 19 and the other is 38 years old. Most women in these situations would probably save themselves, but older women should also be significantly more inclined to save their baby than would younger women.

Like the *pugetensis* sparrows, a young human can have other babies and her best strategy under this situation would therefore be to "cut bait"; by contrast, the older woman, like the *gambelli* sparrow, would likely be sacrificing her only reproductive potential and should be reluctant to do so. Interestingly, most American hospitals take special note if a newborn is the first offspring of an older mother. The presence of other, dependent offspring introduces complications: It should reduce the likelihood of maternal self-sacrifice. It would be adaptive to sacrifice the infant if the failure to do so would endanger the ultimate success of other children to whom the mother is equally related. Our hypothetical mother would therefore be acting in accord with constraints that are ultimately selfish, conferring maximum Darwinian fitness by way of optimizing investment in her offspring.

Of course, such concerns are not limited to women although they would probably be more strongly developed than among men. This is because of the greater female parental investment, the restricted age range of female reproductive potential, and the unique nutritive role of mothers. To some extent, fathers could also be expected to vary their behaviors as a function of fitness considerations, although the uncertainty of genetic relatedness adds a degree of diffuseness to the strategic evaluations. The exemptions of fathers of young children from military service may reflect at least in part an appreciation that such individuals may be less inclined to risk themselves for other causes.

Assuming once again that human behavior has evolved to maximize individual fitness, interactions with others should be patterned with regard to genetic relatedness; see the discussion of kin selection in Chapter 4. This has implications for differential male-female parental strategies. Women have the primary child-care roles in all human societies. Men are significantly less concerned with infants and children. Cultural determinists will claim this to be a function of socialization, but this does not explain the universality of such tendencies. The one commonality shared by Alaskan Eskimos, Australian aborigines, African Bushmen and Wall Street businessmen is their biological heritage; one aspect of that heritage is that males of virtually all animal species must have less confidence in their paternity than females have in their maternity. Females know that they share 50% of their genes with their children, while males must take the female's word for it. It is therefore adaptive for females to invest heavily in the well-being of the children. Males are predicted to be less predisposed in this direction, especially because the solicitous tendencies of women enable such male "irresponsibility" without any decrement in the male's fitness via his presumed offspring. The woman can be counted upon to take care of the kids. She will lactate; he will not.

What will he do? Like the male hoary marmots inhabiting a highly social colony (Chapter 7), the human male can maximize his fitness by interacting with other adults. By competing with other males, he can retain access to his female and also possibly attract additional mates. This line of reasoning thus provides further support for the "biology of the double standard" argument presented above, and it also suggests why women have almost universally found themselves relegated to the nursery while men derive their greatest satisfaction from their jobs. For example, reversion to standard sexual division of labor has even occurred in Israeli *kibbutzim*, despite an overt ideological commitment to behave differently (Tiger and Shepher, 1975).

Like a male red-winged blackbird struggling to defend a territory capable of transforming him from a bachelor to a polygynist, the male human may well have been selected for behaviors that maximize his fitness. According to this line of reasoning, success in male-male competition indicates and/or generates resources ultimately attractive to women for which men will therefore compete. For their part, women then choose these resources because they in turn will be selected to bestow their reproductive favors in accordance with the maximization of *their* fitness. Hypergamy should therefore correlate with polygyny, and it does. Of course, as with other behaviors, modern twentieth century life would have greatly diffused these more primitive, biologically generated tendencies, but they may well persist nonetheless, taking different forms in different cultures.

Such differences in male-female attachment to family versus vocation could derive in part from hormonal differences between the sexes. The maternal hormones prolactin and progesterone are intimately concerned with care of young in most vertebrate species, although there is no clear-cut evidence for such a relationship in *Homo sapiens*. If it did hold, it would constitute *proximate* causation; for the *ultimate*, evolutionary cause, we ask, why should child care be the concern of female hormones? For the same reason that vertebrate aggressiveness is largely the concern of male hormones; for the same reason that sugar is sweet.

Successful parents, insofar as evolution is concerned, are those that become grandparents. Grandparents share one-fourth of their genes with their grandchildren. For the same reason that men and women differ in the confidence of their genetic relatedness to their offspring, grandparents or, rather, grandmothers are certain to be related to the offspring of their daughters, whereas they must rely upon the honor of their daughters-in-law for the offspring of their sons. Grandparents would therefore be predicted to invest more heavily in their daughters than in their daughters-in-law. Why is it usually the mother's parents who help out most when the new baby arrives? On the other hand, patrilocal residence is the most common human living arrangement (van den Berghe, 1975); the bride moves in with the family of her husband. In this manner, the in-laws with less genetic security can oversee their investment. It would also be interesting to know whether pressures for avoiding extramarital intercourse are exerted particularly by in-laws, those with the most to lose.

Of course sociobiology holds that, when parents invest in their offspring, it is because such investment increases the likelihood that they will become grandparents; i.e., evolutionary success, fitness, is ultimately measured by the production of genetically related descendants. The reproductive performance of children should therefore be a major concern of parents. Significantly, young couples often describe parental pressure as a major factor positively influencing their own decision to have children. Of course their parents *naturally* want to become grandparents. The evolutionary perspective of sociobiology suggests why this may be.

We can also make specific predictions. For example, parental concern that each offspring reproduce should to some extent vary inversely with the number of such offspring. The spoiled only child is a well-known phenomenon. We may have been selected to lavish resources on our children in proportion to how much they represent our only chance at long-range reproductive success. This is not to deny parental love in large families, but it is probably no coincidence that the glow of grandparenthood like that of parenthood tends to diminish with repeated kindling.

The sociobiological theory of parent-offspring conflict (Chapter 7 and Trivers, 1974) provides numerous opportunities for predictions concerning human behavior, based once more upon the assumption that indivduals will be selected for the maximization of their inclusive fitness. Children should therefore be selected to demand more parental investment than the parents will be selected to give, and weaning conflict can be expected, for example, with regard to both the amount and duration of nursing. The child wants a lot of milk; where this is necessary for offspring success, the parents also want the child to get it. However, parents want to invest in other offspring as well, and the nursing child can therefore be expected to disagree over the amount and duration of nursing.

Of course offspring would lose fitness if they were too selfish. Excessive gluttony would contribute little if at all to their personal success and, besides, offspring stand to gain fitness from the reproductive success of their siblings in exactly the same sense as the altruistic turkeys and the Tasmanian native hens (see Chapter 4). Genes for extreme selfishness relative to siblings would therefore be selected against by the consequent decline in inclusive fitness, unless the personal reproductive advantage they confer exceeded twice the disadvantage they inflict upon their brothers and sisters (because siblings share half their genes, on the average).

This line of reasoning suggests many interesting predictions. For example, a child's willingness to forego additional parental investment should be to some extent a function of its genetic relatedness to its other siblings. In families or societies where siblings are successively produced by the same parents, children should engage in less parent-offspring conflict with regard to the duration and amount of parental investment than in situations where different, unrelated males successively father the offspring. Similarly, greater offspring-offspring conflict would be expected among half-siblings, the offspring of one man's unrelated wives, than among full siblings.

Furthermore, older parents should experience less parent-offspring conflict than younger parents. This is because the former have less future reproductive potential than the latter and would therefore be less strongly selected for withholding investment from their current offspring. On the other hand, weaning conflict should predictably increase during the development of each child as (1) it becomes increasingly competent to survive without the disputed investment, (2) its increased size requires proportionately more of that investment, and (3) the parent becomes increasingly prepared to invest in subsequent offspring. It is interesting that minimization of parent-offspring conflict over lactation and nursing is institutionalized in many human societies through a postpartum sex taboo until the nursing infant is weaned.

Children would be expected to employ psychological rather than physical tactics in attempting to induce more parental investment than would be in the best interests of the parents themselves. This derives from the obvious physical disadvantage of the offspring. Given that offspring will be selected for the use of such tricks, parents will be selected for the ability to discriminate true need from those demands that would increase offspring fitness at the cost of parental fitness. Previous experience with children might benefit parents in their ability to make this distinction. Therefore experienced parents should be relatively more successful at winning parent-offspring conflicts (Trivers, 1974).

Behaviors characteristic of earlier developmental stages should serve as an effective tactic in generating parental investment, because parent-offspring conflict is expected to be less intense when the offspring are younger. The psychological phenomenon of regression to infantile behavior could therefore have its roots in such a system (Trivers, 1974). Offspring would also be selected for sensitivity to their own needs when these needs relate to their own fitness and to their parents' inclinations concerning future investment in them. Thus, the evolutionary perspective suggests that living things behave with a degree of "enlightened self-interest" not usually attributed to them. For example, several studies have examined the effects of mother-infant separation on the behavior of rhesus monkeys; such research may be of practical significance for understanding and predicting the often-traumatic consequences of forced separation in humans (Bowlby, 1973). Specifically, when rhesus mothers are removed from their infants and then reunited, the infants spend significantly more time huddling with the mother and demanding attention than they did before separation (Hinde and Spencer-Booth, 1971). Furthermore, the more frequent the separations, the greater the infant demands for maternal attention. In contrast, when mother-infant separation occurs by removing the *infant*, rather than the mother, then the infant is less demanding of maternal attention when they are eventually reunited (Hinde and Davies, 1972). This difference may seem puzzling: why should it make such a difference to the infant whether the separation is caused by removal of the mother or of the infant? In either case, the separation is the same. However, an evolutionary perspective suggests an answer. Separation caused by departure of the infant is less likely to indicate maternal negligence than is separation caused by departure of the mother. Therefore, infants should be relatively unconcerned about the former, whereas when they have some reason to doubt their mothers' devotedness to them, it would be adaptive for them to behave in a manner that will reduce the likelihood of such separations occurring in the future. Sociobiology thus offers a rather new perspective on parent-offspring interactions, one that is based upon the notion that each individual is ultimately motivated by considerations of their own evolutionary self-interest: it sheds new light on old problems.

According to the sociobiology of parent-offspring conflict, parents and offspring are expected to disagree over the "behavioral tendencies of the offspring insofar as these tendencies affect related individuals" (Trivers, 1974). For example, parents have an evolutionary interest in preventing their offspring from engaging in a selfish act that harms a full sibling, whereas the offspring are selected for persistence in such behavior so long as the performer benefits from it more than one-half as much as the recipient is harmed, because full siblings have, on the average, one-half of their genes in common. The same potential conflict exists with regard to more distant relatives as well, when the genetic relatedness of parent-to-relative and child-to-relative is assymetric. For example, parents share one-fourth of their genes with their nieces or nephews, while their offspring share only one-eighth of their genes with these same individuals (their cousins), adults should therefore have a greater interest than their offspring in discouraging selfishness and encouraging altruism among cousins.

Psychologists have traditionally viewed early socialization as a process of *enculturation* during which the offspring are simply instructed in the details of their parents' culture for the benefit of all. Clearly, the preceding discussion is greatly at variance with this approach. Given their conflicting evolutionary strategies, we can predict that parents will attempt to induce offspring to behave more altruistically than would be in the children's best interests and that the latter should resist these efforts, tending toward a degree of egoistic behavior that should in turn be resisted by the parents. However, parents have a big edge: Existing adult culture is presumably of some adaptive value, at least as indicated by the fact that its practitioners have survived and succeeded in producing children. In addition, because of their greater experience, parents probably do have something of value to impart to their children. Offspring should therefore be selected for susceptibility to adult teaching. In turn, parents might be expected to take advantage of this vulnerability on the part of their children by exaggerating the significance and importance of their teaching, while using such teaching to further their own manipulation of their offspring. It is therefore not surprising that the prevalent view of socialization, in which children are pictured ideally as passive recipients of parental beneficence, is one that is supported by adults! (Trivers, 1974).

In fact, a great deal of actual parent-offspring conflict may be illuminated by Trivers' approach. Common parental exhortations include going to bed early, studying hard at school, not fighting with siblings, refraining from gambling, drinking, or premarital sex, and learning to share. All these are either directly altruistic (prepare the child for future altruism, increase the likelihood of others reciprocating with altruism toward relatives) or reduce the expenditure of time and/or energy required of the parents. Children often disagree with these parental prescriptions, considering them to be unpleasant, excessively moral, and/or generally a drag. This may well emanate in part from their adaptive unconscious perception that such actions would maximize their parents' fitness rather than their own. Enlightened self-interest may work wonders in generating cooperative children.

According to sociobiologic theory, children are only a special case of our evolved propensity for leaving genetic representation in future generations. We share more genes with our children than with our nieces or nephews, more with the latter than with second cousins, etc. But these differences are quantitative, not qualitative. Parental behaviors should therefore constitute only a special case of generalized concern for the success of our relatives. The family organization in current American society may actually constitute an aberration in terms of our biological history and evolved propensities, in that the nuclear family is restricted to parents and their children. When the children mature, they typically leave the family circle, often in search of educational opportunities, jobs, a spouse, etc., and establish their own nuclear family. Contrast this with the usual pattern of primitive, nonindustrialized societies: Children tend to remain near their parents, and their offspring in turn are brought up in an environment of grandparents, uncles, aunts, cousins, etc.

Assuming that evolution has favored both the expenditure and receipt of kin investment, our present system of relatively isolated nuclear families may well generate substantial stresses upon all concerned. The frustration of grandparents who must content themselves with occasional long-distance telephone calls and eagerly sought Christmas visits may reflect their forced inability to fulfill a major biological satisfaction. Maiden aunts and bachelor uncles are similarly deprived, although they

307

may not consciously recognize it. For their part, parents in isolated nuclear families may well be laboring under stresses that are novel to our biology. Life was certainly not easy for our remote ancestors. But having evolved in small social groups in which genetic relatedness was undoubtedly quite high, we as parents probably enjoyed considerable assistance from our relatives when it came to the obligations of child rearing. Such assistance was beneficial to all concerned. Older individuals were experienced and knowledgable in child care, while juveniles and adolescents profited by the training. Finally, the dependent children themselves doubtless gained by the arrangement. Modern domestic strains can be somewhat alleviated by babysitters, day-care centers, and, possibly, communal living arrangements, but for the most part our turning from biology may have robbed child rearing of much of the serenity it once possessed.

In addition, extended-family societies provide an excellent opportunity for evaluating the applicability of inclusive fitness theory to human social behavior. Most of them have unilineal descent; i.e., they are either patrilineal, tracing kin relationships through the father's line, or matrilineal, through the mother's. Given that genetic relatedness is equivalent in both cases, the question is whether the cultural rule of ignoring all but one line of descent predicts altrusitic behavior better than does actual biological relatedness. The evidence is simply not in, and the question remains open.

Altruism and Such

We have discussed the biology of altruism among nonhuman animals (Chapter 4), specifically excluding any consideration of internal motivation or cognition. In contrast, the social science literature on human altruism revolves almost entirely around such notions. In this brief section let us apply a sociobiological approach to human altruism, suggesting possible adaptive predispositions in *Homo sapiens*. Once again, this is not to deny a role for learning or social tradition in mediating such behavior;

factors of this sort are entirely compatible with underlying genetically influenced tendencies as well. The value of sociobiology's evolutionary approach is that it allows predictions of possible behavioral universals or at least a common substructure rooted in our biology.

The theory of kin selection has great potential relevance to human behavior. In fact, it suggests a coherent theory for the biology of nepotism. It provides a straightforward answer to the puzzle of why we tend to favor our relatives. Sociobiology's prediction is that we should be selected to show altruism toward others in direct proportion to how closely related they are to us genetically. Insofar as the Central Theorem of Sociobiology holds for us, the relationship $k > 1/r$ (Chapter 4) should hold for human altruism. We should be willing to suffer greater risks in aiding individuals who are more closely related and should withhold aid to more distantly related individuals, unless the risk to our personal, Darwinian fitness is also proportionately lower. Similarly, we should require that a distant relative be in greater need, derive a higher benefit from our act, in order for us to render the same assistance that we would dispense to a closer relative in less dire straits. Anxious parents may inquire eagerly as to whether their children need more money at college, whereas, in general, assistance to a second cousin might be forthcoming only in greatest emergency and even then only after it is requested.

Assuming equivalent genetic relatedness to all recipients (equal r), our willingness to disperse altruism should vary directly with the benefit in fitness to the recipient and inversely with the cost in fitness to the altruist. Appeals for the needy emphasize how much good can be done with so little cost: "Only one dollar provides milk for a month". Similarly, costs are lowest for the potential altruist when his or her personal reproductive potential is lowest; there is little to lose. Older, post-reproductive people should therefore be particularly inclined to assist others. If they have many potential beneficiaries that are also relatives, their attention would be especially directed toward them. The need to help or to be needed should be strong in all *Homo sapiens*, and feeling useless should be painful indeed. On the other hand, there should be a sweetness to life when it accords with the adaptive wisdom of evolution.

The literature of anthropology is crammed with the complex, involved patterns of human kinship systems. But this is not just an aberration of anthropologists; it also reflects the universal concern of people. As a species we are obsessed with the identification of genealogies and blood relationships. Every human language devotes numerous terms to the recognition of individuals in accordance with their genetic relatedness: grandparents, uncles, cousins, etc. In so doing, each of us identifies ourselves in a unique, personalized network of evolutionary concern, most densely interwoven among our immediate family and getting progressively thinner on the social periphery. Kinship systems clearly have enormous influence upon human social interactions, although, surprisingly, anthropologists have been unable to agree as to the reason for the universality of kinship as an organizing principle. Instead, they have concerned themselves with identification of the *ways* in which kinship considerations serve to organize human social interactions; i.e., they have taken a structural rather than a functional approach.

Nepotism is very real and, significantly, very reasonable to most people. An analysis of human bias toward kin, as revealed, for example, in wills, would almost certainly support the evolutionary prediction for such a tendency. On the other hand, of course we favor our close relatives "because" we often spend considerable time with them. An evolutionary approach suggests the inquiry: Why do we spend time with our relatives and, further, why do we love as a result? *If we admit to the possibility that human behavior has been selected to maximize inclusive fitness,* then our preoccupation with genetic relatedness and our responses to it are certainly no surprise. In fact anything else would require some explanation.

Genetic relatedness often declines dramatically beyond the boundaries of a social group (Hamilton, 1975) and, significantly, aggressiveness increases in turn. Hostility toward outsiders is characteristic of both human and nonhuman animals. Physical similarity is also a function of genetic relatedness, and human racial prejudice, directed against individuals that

310

look *different*, could have its roots in this tendency to distinguish in-group from out-group (P. Greene, ms. in preparation). Human populations have long been dispersed over a wide geographical area. As local populations adapted to their particular environments, differentiation of physical characteristics occurred, reflecting these local adaptations. There are many theories for the adaptive value of human skin coloration, but we shall not be concerned with them at present; see W. J. Hamilton (1973) for a review. Regardless of the cause of such distinctiveness, it certainly occurred. Given the evolution of differences and the fact that such differences have been associated with greater genetic similarity within groups than between groups, it seems plausible that a degree of antagonism, the converse of altruism, would occur when representatives of different groups meet. Thus, our prejudiced comments relative to others tend to emphasize overt characteristics that distinguish them from ourselves: honkies, niggers, chinks, etc. Clearly, this suggestion of a possible evolutionary basis for human racial prejudice is not intended to legitimize it, just to indicate why it may occur: Behavior patterns that may have been adaptive under biological conditions are inappropriate and even dangerous under the cultural innovations of today.

Of course human altruism is not necessarily reserved exclusively for relatives. We should bear in mind that the cultural framework of modern Western society represents the human condition during less than 1% of its evolutionary history. Whatever the influence of biology upon our behavior, this imprint was established during the previous 99% of our existence as a species. We cannot reconstruct the formative influences with certainty, although many anthropologists justify their studies of primitive human societies and of nonhuman primates by pointing to the relevance of such research to an understanding of pre-historic human behavior.

Biological evolution is a slow process. *Individuals* do not evolve although natural selection operates by differential reproduction of individuals: each living thing is limited to the genetic make-up it inherits. *Populations* evolve; and because evolution proceeds by the differential reproduction of indi-

311

PLATE 10.1 A SMALL GROUP OF MUSK OXEN IN A LOOSE DEFENSIVE PATTERN.
Behavior of this sort is highly adaptive against wolves, but not against human hunters. However, since hunters armed with guns are very recent insofar as the evolution of musk ox behavior is concerned, these animals have not yet evolved a different and more adaptive response. (PHOTO BY C. SHANK)

viduals comprising a population, biological change of this sort is restricted by the genetic possibilities available and by the time necessary to produce a new generation as well as by the selective pressures and the heritability of the trait(s) involved. Hundreds and more often thousands of generations are required for natural selection to produce significant evolutionary change. However, drastic cultural change has often occurred during a single human lifetime. For this reason, human behavior patterns may sometimes seem curiously nonadaptive today simply because they are being assessed in an environment far different from the one in which they evolved.

When musk oxen encounter a predator, they form a circle with the formidable horns of the adult males facing out and the vulnerable females and young protected in the center. This is an effective and therefore adaptive response to wolves, their natural predators. But it is a suicidal strategy against humans armed with guns (Plate 10.1). Similarly, the adoption of an unrelated child seems curiously nonadaptive and a potential example of the human tendency to perform true altruism, counter to the predictions of evolutionary theory. But consider that the greatest part of our evolutionary past was probably spent in small social groups (Lee and Devore, 1968). Therefore orphans were likely related to the older individuals who adopted them. These altruistic adults could therefore have been acting to increase their own inclusive fitness: such behavior would accordingly have been favored by kin selection. Note that maximization of fitness need *not* have been the conscious intent of early human adopters.

Sociobiological theory would further predict that adults with the most to gain and the least to lose would be the most eager adopters, and certainly this is true in the United States where childless couples are the predominant adopters. In addition, solicitude toward an orphan may ensure eventual assistance by that adopted child toward relatives of the adopting adults and hence be selected by the mechanism of reciprocal altruism. Similarly, adopting adults may also receive additional benefits from other individuals in the society whose gratitude at their behavior may derive in part from their own relatedness to the orphan and, hence their own interest in its welfare. In present-day Western society, the chances of genetic relatedness among adopters and adoptees are virtually zero, but the behavior may persist, like the outmoded circle defense of musk oxen, because our biology has not had sufficient time to accommodate to culture's rapid development. The care of young children is something most people find sweet.

Alternatively or additionally, we may behave altruistically toward nonrelatives largely because of societal exhortations and conditioning. Let us assume that maximization of fitness will have as its primary effect selection for various forms of selfish behaviors. In that case, religious, moral, and other forms of cultural influence upon human behavior may reflect the necessity of restricting these tendencies in order for complex societies to function effectively (Campbell, 1975). This fascinating idea suggests that there may be a certain adaptive wisdom in such injunctions as the Ten Commandments or the Bhagavad-Gita. On the other hand, it also suggests that society has an existence and interest independent of its constituent members. Animal social systems are most parsimoniously interpreted as the outcome of each individual making the best decision it can for itself. If we adopt a similar perspective regarding humans then adherence to social convention may actually represent maximization of fitness for each individual (Greene and Barash, 1976).

Back to altruism: reciprocity provides a cogent possibility for genetically influenced altruism among nonrelatives (Chapter 4 and Trivers, 1971). In fact, the rather stringent conditions required for the evolution of reciprocity make *Homo sapiens* the most likely candidate for such a process. However, an individual within such a system will invariably benefit if he or she can receive aid from others while refusing to reciprocate when the tables are turned. Cheaters thereby receive an evolutionary boost from others without incurring any risk; such strategies would be more fit than inclinations toward true reciprocity, and therefore should prevent the evolution of such behavior. On the other hand, evolution of this sort would also select strongly for ability to identify cheaters and discriminate against them (Trivers, 1971). Our great concern with the evaluation of each other's character, trustworthiness, and motives behind one's actions takes on new potential significance when viewed with this perspective.

The theory of reciprocity suggests that altruism will characterize communities in proportion to their composition of individuals who know each other. To some extent, this should distinguish small towns from large cities and may relate to the infamous reluctance to get involved, so characteristic of the latter. Inhabitants of a small town would be unlikely to ignore a murder or step over a body on the sidewalk. In addition, the theory suggests tendencies for altruism in proportion to how sedentary the population is; the altruist has a reasonable expectation that his beneficiary will have the opportunity to reciprocate. This would reinforce greater altruism in small, rural communities where physical mobility is reduced. Reciprocal altruism would also be favored when the benefit to the recipient is high, i.e., the need is great, but might occasionally be countered by a reduced tendency toward such behavior when the cost to the altruist is also high, i.e., the act is risky or requires expenditure of a scarce resource. Under the latter circumstances, kin selection should predominate over reciprocity, and, as the situation deteriorates further, the scope of kin concern should narrow to one's own offspring, i.e., selfishness.

Finally, altruism toward unrelated individuals should be most adaptive when directed toward those with the greatest likelihood of reciprocating effectively, thereby increasing the inclusive fitness of the altruist. Therefore preferred beneficiaries should have access to resources that they could subsequently bestow upon the altruist and/or his relatives: Money, promotions, etc. Discounting the resource factor, altruists should accordingly be more fit if they direct their attention toward individuals who are more likely to live long and thus ultimately be able to benefit the altruist; we should favor children over old people. National indifference toward senior citizens may be relevant here, as well as the reluctance of physicians to specialize in geriatrics as compared to the popularity of pediatrics. And why does the heart-rending photo of the starving Biafran always feature a young child?

Reciprocity theory per se is not novel in the social sciences and, indeed, notions of enlightened self-interest go back at least to Machiavelli and Hobbes (the 16th and 17th centuries). It is therefore certainly possible that the above considerations represent gratuitous biologizing. Indeed, this may be true of much of sociobiologic theory, especially when applied to human behavior. However, at this primitive stage in our understanding of *Homo sapiens* and our equally primitive understanding of sociobiology, it seems entirely appropriate that we explore all possible avenues, so long as we do so with appreciation of their weaknesses, as well as their potential strengths.

Marshall Sahlins (1966), an anthropologist, has presented a diagram depicting the patterns of altruistic interactions among individuals in nonindustrialized human societies (FIG. 10.2). His findings are particularly significant because they were developed without reference to sociobiological theory, and yet they coincide exactly with the prediction that behavior will tend to maximize inclusive fitness. Thus, as we progress toward the

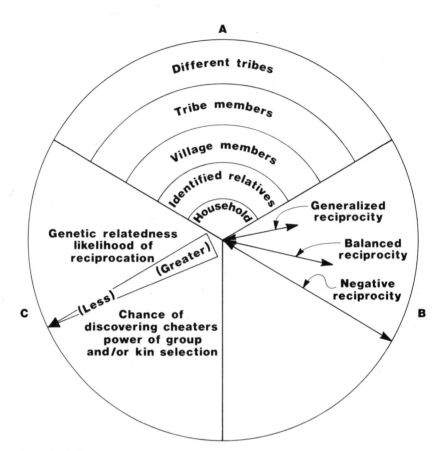

FIGURE 10.2. CORRELATION OF (A) SOCIAL RELATIONSHIPS, (B) PATTERNS OF RECIPROCITY, AND (C) FACTORS SELECTING FOR ALTRUISM IN PRIMITIVE HUMAN CULTURES. The first two phenomena were identified by Sahlins (1965); they accord with sociobiologic predictions based on genetic relatedness as well as the conditions required for the evolution of a reciprocal altruism. (*Modified slightly from Alexander, 1975.*)

outer concentric circles, genetic relatedness among individuals declines progressively, exactly equivalent to the networks of evolutionary concern described above. Altruism should therefore be most intense within the center, declining progressively toward the outer rings. And it is.

The likelihood of reciprocity can also be overlaid upon this system; its direction parallels that of kin selection. The closer the genetic relationship, the greater the frequency and duration of physical association and therefore the stronger the selection via reciprocity as well. Similarly, the more distant the relationship, the farther out on Sahlins' circle (FIG. 10.2), the lower the probability that opportunities for reciprocity will ever occur. Why do a good deed for a total stranger if it is very unlikey that he or she will ever return the favor? The great appeal but nearly universal disregard of Christian ethics is suggestive evidence in this regard. The Golden Rule works best among relatives, friends, and sometimes, acquaintances. Beyond this, the penalty for cheating diminishes steadily with decrease in genetic relatedness along with reductions in the likelihood that cheaters will be brought to account for their misdeeds.

Accordingly, interactions between tribes tend to be exploitative, even to the point of stealing and war. Many nonindustrialized societies have no ethnic term to designate themselves other than the word for "people". Western societies are not exempt and, in fact, given the large numbers of nonrelatives and transients with which Westerns typically deal, it may well be that members of the lonely crowd are particularly vulnerable to selfish exploitation or, at least, indifference by others.

Other Predictions, Universals, and Levels of Complexity

The predictions discussed above certainly do not exhaust the capabilities of human sociobiology. They are just a beginning. Predictive power will probably be greatest for behaviors that are minimally removed from genetic influence, such basic phenomena as reproducing, eating, resting, sharing, and fighting. Psychology in particular should profit from this evolutionary approach, and sociology and anthropology to a somewhat lesser extent when they deal with more complex, emergent phenomena. Political science and economics may also derive some insights, but as concern moves to higher-order processes—the interactions of diffuse, multipurpose social structures with each other—we get increasingly removed from biology. However, the personal motivations of the individuals involved in such systems may still relate to the maximization of their own inclusive fitness. Evolution probably offers more insight into the behavior of individuals, than of their creations.

My concern in this chapter was to generate predictions based upon the ultimate generative force of evolution; natural selection acting to maximize the inclusive fitness of individuals. Of course, there are other possibilities. All of recorded human history represents only a tiny fraction of our evolutionary past. Accordingly, it may also be productive to try a more historical approach: Estimate the ecological factors impinging upon the human species during its formative years, presuppose certain adaptive behavioral responses to these conditions, and then assess the extent to which such evolutionary baggage influences *Homo sapiens* in the modern world. Statements about dominance orders and territorial behavior would be especially appropriate to this technique (van den Berghe, 1974). Other cultural traits such as religion, artistic expression, division of labor, and educational practices may also be susceptible to such analysis, perhaps especially if it is used in conjunction with the philosopher's stone of fitness considerations.

The Conflict between Culture and Biology in Human Affairs

The human species lives a schizophrenic life. We are the only uncomfortable animal, strangely out of place in our own environment. This may well be due to our unique, dichotomous nature as both biological and cultural creatures. We are the products of biological evolution: A slow, natural process that proceeds by the differential reproduction of individuals and, hence, the gradual replacement of genes in a population. But we are also experiencing cultural evolution, an incredibly rapid progression of events that occurs even within the lifespan of a single individual. In a sense, we are all time travelers, with one foot in our biological, evolutionary past and the other balanced precariously on our rampaging cultural present. This basic discrepancy of human life may underlie most of our problems today. Adequate discussion of this question clearly requires a separate treatment. Therefore, the remainder of this section will attempt only to outline some of the basic issues.

Of course, there is a sense in which biology and culture are not antagonistic. Thus, culture is a major biological adaptation of *Homo sapiens*; and certainly, we possess a biologically-evolved *capacity* for culture. But cultural evolution is very different from biological evolution. Culture is essentially Lamarckian (see Chapter 2) in that characteristics acquired via experience or discovery are transmitted directly to others and thereby immediately incorporated into the cultural phenotype. As a result, cultural evolution moves at incredible speed, such that biological evolution may very well never catch up. This is not to say that biological evolution has stopped in our species—so long as some individuals produce more offspring than others, natural selection is taking place. We may soon be selecting for resistance to DDT and carbon monoxide, or perhaps ability to avoid auto accidents. One thousand years from now, humans will doubtless seem very different from what they are today, but it is probably a good bet that there will have been little biological change in that time: Even a thousand years is very short in terms of biological evolution. But just imagine the cultural changes during the same period! The discrepancy between culture and biology will probably be even greater then, although there is every reason to believe that it is uncomfortably large even now.

For example, animals that possess the physical equipment enabling them to kill others of the same species usually refrain from doing so (Lorenz, 1952). Thus, intraspecific killing is very rare in wolves, rattlesnakes, eagles, etc. Arabian oryxes use their rapierlike horns to good effect against lions, their natural predators; in contrast, when one oryx fights another, the horns are braced against each other, but the sharp tips are never used. In fact, the horns are outfitted with ridges helping to ensure that the animals will not slip apart while they test each other's strength (Plate 10.2). Similarly, giraffes have long legs and sharp hooves, good weapons when used against predators, as they often are. When fighting another giraffe, however, they refrain from kicking; rather, they butt heads, employing their

PLATE 10.2 TWO MALE ARABIAN ORYX ANTELOPE IN RITUALIZED COMBAT.
They refrain from using their potentially lethal horns against each other, re-
serving them instead for their predators, such as lions. (SAN DIEGO ZOO
PHOTO)

harmless stubby horns. The list is extensive: Rattlesnakes never
bite each other while fighting; deer and elk have branching
antlers that interlock with their opponent's, although sharp
pointed antlers would be far more deadly (Plate 10.3). In fact,
male deer have actually been observed to refrain from attacking the unprotected flank of a rival, waiting instead until they
can lock antlers (Eibl-Eibegfeldt, 1961).

In most cases, then, animals possessing the biologically
evolved capacity for killing a conspecific often also possess
biologically evolved inhibitions against doing so. Such inhibitions are often reflected in highly ritualized and stereotyped
display behaviors that can be interpreted as resulting from
selection operating to maximize the fitness of individuals
(Chapter 8). In contrast, animals lacking the capacity to kill also
lack inhibitions against doing so. And we must include *Homo sa-*

piens within that group. Thus, it is very difficult for a naked, unarmed human being to kill another; we simply do not possess the appropriate biological equipment. Accordingly, it seems likely that we have not evolved inhibitions against killing; there was no need. But within the last few thousand years we have provided ourselves, through cultural evolution, with progressively more efficient means of killing. We boast knives, bows and arrows, guns, and now, nuclear weapons. All these "advances" have occurred within a moment's time as biological evolution reckons these things. We have not had the time needed to evolve the appropriate inhibitions. Furthermore, even if we did, the invention of efficient long-range killing devices would render such behaviors largely irrelevant anyhow: The bombardier at 10,000 feet could not even see the potential subordination behaviors of the inhabitants below.

The result of this discrepancy between our biology and our culture is that we are literally in danger of destroying ourselves (Lorenz, 1966). A similar analysis reveals that in fact most gen-

PLATE 10.3 TWO MALE PERE DAVID DEER LOCKING ANTLERS IN RITUALIZED BATTLE FOR DOMINANCE. Such animals refrain from using their weaponry to kill an opponent. (PHOTO BY NEW YORK ZOOLOGICAL SOCIETY)

eral human problems may well derive from this fundamental incongruence of human life. For a less disturbing example: Earlier in this chapter we considered why sugar is sweet. Our adaptive fondness for sugars has actually become a major nuisance now that cultural evolution has discovered that it can pander to this biologically evolved trait. Through culture we provide ourselves with the essentially nonnutritive cakes, cookies, chocolates, and assorted effluvia of the sweet-tooth industry. We know its not good for us, but we are still biological creatures and can't easily resist our fondness for sugars.

A similar argument applies to another major human health problem, the various ailments associated with insufficient exercise, especially heart disease. Most animals are lazy: The lethargic zoo lion is probably less miserably bored than is the average spectator sympathizing with his "plight". Free-living animals normally expend a great deal of energy just staying alive. This was almost certainly true of our hominid ancestors attempting to find food and avoid predators on the vast African savannahs during the Pleistocene epoch. Under such circumstances, the *avoidance* of exercise would be adaptive, since it would involve a saving of precious calories. However, such opportunities would be rare, so the human body probably evolved with both the expectation (i.e., biological need) for regular, vigorous exercise and an adaptive propensity to avoid it whenever possible.

Now of course we have automobiles, telephones, and elevators and nearly everyone practices a sedentary occupation; unless we actively seek physical exercise, most of us don't get any, and we are disinclined to seek it. Again, the whole thing is not unlike the musk oxen forming a circle when hunted: We are employing a strategy that our genes tell us is adaptive but that modern culture has rendered nearly suicidal. To make matters worse, many of us also appear to have inherited a fondness for rich, fatty foods; game animals tend to be lean, and therefore animal fat would be uncommon, high in calories, and a valuable and adaptive treat. Accordingly, cultural evolution now permits us to combine less exercise with an overabundance of cholesterol. The conflict between cultural and biological evolution may at least be profitable for confectioners, French chefs, health spa operators, dentists, cardiologists . . . and undertakers.

There are many other examples: We are unique among animals in the extent to which we pollute, despoil, and thereby endanger our own environments. Of course, our capacity for environmental destruction is largely a function of our cultural advances of the past few thousand years, in particular, the past hundred or so. As with our relative indifference to intraspecific killing, our culturally evolved capacity for destruction is comparatively unhindered by any biologically evolved inhibitions. On a more immediate level, it is interesting that *Homo sapiens* with all its vaunted brain power is harder to toilet train than a dog or cat. The same holds for all other primates (one reason monkeys make terrible pets), suggesting that in the absence of a permanent den, inhibitions against fouling one's nest may be undeveloped. Besides, fecal material does not remain long in a monkey's arboreal world anyhow, so there would be no special adaptive value to such concern. So, we're blithely on our way, free of biologically generated inhibitions, fouling our world nest on an unprecedented scale.

Another major source of our current environmental dilemma is the sheer numbers of our species. Overpopulation is one of humanity's most serious problems and it also derives from the conflict between biological and cultural evolution. In the long run, the population level of any species is determined by the birth rate minus the death rate. Birth rate is a consequence of our adaptive, biologically evolved tendencies for reproducing. For all animals, and pre-technological *Homo sapiens* as well, death rates were also determined biologically. But cultural evolution has produced dramatic changes in the human situation; through agriculture, food distribution, control of epidemics, etc. we have intervened culturally and have greatly reduced the death rate. However there has been relatively little compensatory change in the birth rate, our remaining biological heritage. The result is comparable to what happens when we plug up the drain in a sink, leaving the faucets running. Once disrupted in this way, the equilibrium can only be established

323

again by either increasing the death rate (unstopping the drain) or reducing the birth rate (turning down the faucet). Alternatively, we could have a real mess. Having intervened culturally in an otherwise delicately balanced biological system, we have no choice but to employ culture further in hopes of redressing the balance. An added problem is that the current situation of high infant survival and abundant resources seems ideal to our genetically influenced wants. Indeed, our deep, evolutionary "sweet receptors" predominate over our rationality, and herein lies a serious threat to ourselves as a species.

The sheer size and complexity of modern society is a culturally evolved phenomenon, novel to our biological experience. So is technology. Alienation, disillusionment, the whole range of characteristic disaffections with modern life may well have their roots in the dissonance between the relative simplicity of our biological past and the diffuse complexity of our culturally mediated present. Daily, we interact with people we do not know and often do not even see. From our flush toilet to our automobile, we rely constantly upon devices we understand dimly or not at all. We read the morning newspaper and feel the anxiety of being influenced by people and events over which we have no control. The precise consequences of all this cannot be predicted, but we should not be surprised at the renewed general interest in back-packing, natural foods, natural childbirth, and the glorification of simplicity and experience over artifice and rationality.

There are no easy solutions to our dilemma as a species. Certainly, there can be no solutions at all unless we first recognize the problem. We cannot simply wipe out our meddlesome culture and return to glorious savagery; it wouldn't seem very glorious anyhow. No, we are riding a fearsome tiger on a one-way trip. Maybe we can slow it down a bit, at least so we don't fall off and get eaten, and perhaps some day we might even succeed in mastering the beast after all.

Glossary

Adaptation (noun) Any characteristic of a living thing that contributes to its fitness.
(verb) The evolutionary process of acquiring adaptations.

Agonistic Behavior Aggressive behaviors such as threatening and fighting, as well as defensive responses to aggression, such as submitting and fleeing.

Allele One of several alternative forms of the same gene; thus, ABO blood types in humans are formed by three alleles, A, B and O, any two of which are present in an individual.

Alloparent An individual other than the biological mother or father who regularly performs parental behaviors.

Altruism Behavior that reduces the Darwinian fitness of the performing individual while increasing that of the recipient. Biological theories for the evolution of altruism involve primarily group selection, kin selection, and parental manipulation. By this definition, reciprocity does not involve altruism, since it is essentially selfish.

Analogy A similarity in behavior, structure, or other phenotype due to evolution selecting for such similarity rather than to common ancestry. For example, the wings of a bird and of an insect are analogous. (*Contrast with* **homology**)

Androgens Any of a class of hormones particularly associated with male sexual behavior. Testosterone is the best known.

Assortative Mating The tendency of an organism to mate another of its kind. One evolutionary consequence of such mating is a change in the frequency of genotypes, but not of genes.

Caste A group of individuals within a particular species that are physically and behaviorally distinct from others and that are relatively specialized in their behavior. Caste evolution is most extreme in the **eusocial** insects, in which nonreproductive castes are found.

325

Cerebral Cortex Anterior brain region concerned especially with higher mental function.

Character Displacement The evolutionary process wherein two species diverge in areas where they overlap geographically.

Chromosome Structure within a cell nucleus that contains the primary genetic material (DNA) of an organism.

Cichlid A member of a family of fishes (Cichlidae) that exhibits elaborate courtship and reproductive behaviors, often including carrying the eggs and young within the mouth of one parent.

Clutch Number of eggs laid by a female at one time.

Conspecific A member of the same species.

Contact Species Species whose members typically spend considerable time touching each other in contrast to **distance species**.

Convergence The evolutionary process wherein two species become increasingly more similar.

Darling Effect The stimulation of reproductive activity by the presence of other individuals. The typical example of the Darling effect is the greater reproductive success of gulls at large as opposed to small colonies.

Deoxyribosenucleic Acid (DNA) A complex organic molecule that is the stuff of which genes are made.

Diploid Possessing two sets of chromosomes. (Contrast with **haploid**)

Directional Selection Selection favoring individuals at one extreme of variation causing the distribution of the trait in question to shift in that direction, in succeeding generations.

Display A behavior or series of behaviors that have been modified by evolution (through **ritualization**) to serve communication. Most displays are concerned with reproductive or agonistic behaviors.

Distance Species Species whose members are more solitary and separate in contrast to **contact species**.

Dominance Hierarchy A social pattern in which individuals are associated with each other in predictable ways relative to access to resources. Hierarchies depend upon agonistic behavior, although the actual behaviors by which they are maintained may be very subtle. The end result, however, is that some individuals have greater access to resources than do others.

Drosophila Fruit flies; these animals are especially popular as subjects for scientific research in genetics as well as behavior.

Epideictic Behavior A term coined by V. C. Wynne-Edwards referring to the displays whereby individuals inform each other as to their local population size. As a result of such behavior, individuals are supposed to adjust their reproductive activity to avoid over-population.

Ergonomics Patterns in the division of labor within a species. A theory of caste ergonomics was proposed by E. O. Wilson for certain insects. There is as yet no ergonomic theory for vertebrates.

Ethogram A detailed description of the behavior patterns of a particular species, usually concerning single individuals or interactions between pairs of individuals. (*Contrast with* **sociogram**)

Ethology The study of animal behavior conducted primarily by zoologists that emphasizes the observation of animals in a natural or semi-natural setting. Historically, ethology has relied heavily on the theory that certain behaviors have a large genetic component.

Eusociality The presence of nonreproductive castes. Eusociality is especially prominent among bees, ants, and wasps, where it has posed a significant problem for evolutionary theory. Recently, W. D. Hamilton interpreted such altruistic reproductive restraint through the theory of kin selection.

Extinction The evolutionary termination of a species caused by death and failure to reproduce.

Fitness A measure of evolutionary success that applies to genes, traits, individuals, or populations. More precisely, fitness is a number that, when multiplied by the proportion of members in one generation, gives the proportion of representatives in the next.

Gamete A cell (either egg or sperm) specialized to achieve sexual reproduction by combining with another gamete produced by a member of the opposite sex. Females produce a relatively small number of gametes, in each of which they invest heavily; males produce a large number of gametes and typically do not invest as much in them.

Gene Frequency The proportion of a particular gene (more precisely, allele) in a population.

Genetic Assimilation An evolutionary process by which the genes responsible for producing a particular phenotype are accumulated within individuals because of selection for those individuals possessing a certain frequency of those genes. In this way, natural selection can eventually produce a given phenotype by initially favoring those individuals who respond in a particular manner to a given environmental factor. The phenomenon appears Lamarckian but is not. (*Compare with* **Lamarckism**)

Genetic Drift Changes in the distribution of genes within a population due to random effects attributable to small sample size.

Genetic drift operates without regard to adaptive significance; it is less potent as populations are larger and as natural selection operates more strongly.

Genotype The genetic composition of an individual. *(Compare with* **phenotype***)*

Genus A grouping of closely-related species. Domestic dogs, coyotes and wolves are each considered different species, but are members of the same genus *(Canis).*

Group Selection Selection operating via the differential reproduction of groups of individuals as opposed to the individuals themselves (i.e., **natural selection**). Group selection has been proposed as a mechanism for the evolution of altruism.

Haploid Possessing only one set of each chromosome. In complex animals, adults are **diploid** and produce **haploid** gametes, thereby restoring the diploid number upon fertilization.

Hardy-Weinberg Law The statistical relationship between allelic and genotypic frequencies within a population. The equilibrium that results is a system in which evolution ceases. Accordingly, the study of evolution is the study of those factors that disrupt Hardy-Weinberg equilibrium, especially natural selection.

Hemoglobin One of several blood proteins, particularly important for transporting oxygen. Normal hemoglobin is distinct from *sickle-cell hemoglobin,* which is maintained in human populations through **heterosis**.

Heritability The proportion of variance in a trait (within a population) that is attributable to genetic variance among the individuals comprising the population. Heritability may also be defined as the response to selection divided by the selection differential, i.e., the extent to which selection successfully changes the distribution of a trait in a population.

Heterosis A marked vigor occurring when **heterozygotes** are more fit than either **homozygote**. Also known as hybrid vigor, heterosis is often responsible for the maintenance of considerable genetic variation within a population, since it tends to perpetuate both homozygotes.

Heterozygote An individual possessing two different **alleles** with regard to any genetically influenced trait. *(Compare with* **homozygote***)*

Home Range The area traversed by an animal performing its normal activities. Home ranges are not defended and may overlap extensively. *(Contrast with* **territory***)*

Homology A similarity based on common evolutionary descent that may or may not reflect functional similarity. For example, the arms of a gorilla, the front legs of a dog, and the wings of a bird are all **homologous** structures. *(Compare with* **analogy***)*

Homozygote An individual possessing two identical **alleles** with regard to any genetically influenced trait. *(Compare with* **heterozygote***)*

Hormones Chemical messengers. Substances produced by specialized glands that influence other parts of an animal's body. *(Contrast with* **pheromones***)*

Hypergamy "Marrying up"—the tendency of members of one sex to preferentially mate with members of the other who are somehow superior to them.

Inclusive Fitness The sum of an individual's fitness as measured by personal reproductive success and that of relatives, with those relatives devalued in proportion to their genetic distance, i.e., as they share fewer genes. Inclusive fitness is the accumulated consequences of kin selection for any individual.

K-Selection Selection that favors competitive ability within populations thereby inhibiting rapid growth. K-Selection generally produces smaller numbers of offspring with greater investment in each. *(Contrast with* **r-Selection***)*

Kin Selection Selection of genes because of their effect in favoring the reproductive success of relatives other than offspring. Kin selection is especially important in that it suggests a biological basis for altruism, so long as the beneficiary is related to the altruist.

Lamarckism A theory of organic evolution which asserts that environmental changes cause structural changes in organisms that are transmitted to offspring; the inheritance of acquired characteristics.

Lek An area where animals assemble for communal courtship displays.

Matrilineal Tracing kinship lines through the mother's family. *(Contrast with* **patrilineal***)*

Mimicry Mimesis, imitation. A mimetic species typically has been selected for resemblance to another organism or to a natural object common to its environment (the *model*) because the imitation secures the specie a selective advantage (as protection from predation).

Monogamy A reproductive system in which one male and one female form the primary breeding unit.

Mutation Any inheritable change in the genetic make-up of an organism. Mutations are generally rare, small in their effect, and harmful; occasionally, however, they are beneficial, in which case they are selected.

Natural Selection Differential reproduction of individuals; the tendency for some individuals to produce more successful offspring than others. Natural selection is generally acknowledged to be the primary force responsible for evolution.

Nepotism Favoring of one's relatives. Kin selection provides a possible biological basis for this universal characteristic.

Niche The sum total of all physical and biological requirements for a species. Most ecologists now agree that different species occupy different niches.

Nuclear Family A social unit consisting of mother, father and immediate offspring, excluding other relatives such as aunts, uncles, cousins, grandchildren, etc.

Ontogeny The development or course of development of the individual organism. (*Contrast with* **phylogeny**)

Pair Bond A behavioral affiliation between an adult male and an adult female associated with reproduction and especially characteristic of monogamous species. However, pair bonds may also be of brief duration.

Paradigm An organizing conceptual approach and/or principle. Evolution by natural selection is the fundamental paradigm of sociobiology.

Parental Investment Behavior of a parent toward its offspring that increases the chances of that offspring's reproductive success at the cost of the parent's investment in other offspring.

Parental Manipulation A proposed mechanism for the evolution of altruism whereby parents are selected for the production of offspring who behave toward their siblings in a manner that enhances the parent's fitness but at a cost in fitness to the manipulated offspring.

Patrilineal Tracing kinship lines through the father's family. (*Contrast with* **matrilineal**)

Phenotypes Any physical characteristic of an organism whether anatomical, physiological, behavioral, etc. All phenotypes are produced by the interaction of **genotypes** and environment for each individual.

Pheromones Chemicals produced by one organism that influence the behavior of another organism. These "external hormones" are present in urine or feces, and/or they are produced by highly specialized glands.

Phylogenetic Inertia A term coined by E.O. Wilson referring to the set of genetically influenced characteristics of a species making future evolutionary changes more or less likely.

Phylogeny The evolutionary history of a genetically related group of organisms. Human phylogeny is traced through Australopothecines, early primates, certain mammal-like reptiles, etc. Much of ethology has been concerned with the elucidation of behavioral phylogenies. (*Contrast with* **ontogeny**)

Pleistocene A geologic era dating about one million years ago; during this era humans appeared in the fossil record.

Polyandry A reproductive system in which one female mates with more than one male. In many such cases, the male is then the primary parent. (*Contrast with* **polygyny**)

Polygamy A reproductive system in which a single adult member of one sex (either male or female) mates with several members of the opposite sex. Polygamous systems may be either **polyandrous** or **polygynous**.

Polygyny A reproductive system in which one male mates with more than one female. This is the most common system among vertebrates; the female is typically the primary parent.

Polymorphism The occurrence of distinct forms, or discontinuities, as opposed to smooth intergradations between forms. Polymorphisms are of great evolutionary interest; they are often maintained by **heterosis**, when rare types are favored, or by discontinuities in the environment.

Predator Pressure The influence of predators upon their prey. In the short term, predator pressure modifies the population biology of prey species; in the long term, it selects for various anti-predator strategies.

Presenting In non-human primates, a behavior indicating subordination, in which an animal places itself on all fours and elevates its rear end toward another.

Proximate Causation The immediate factors responsible for a particular response: internal physiology, previous experience, factors in the environment, etc. (*Contrast with* **ultimate causation**)

r-Selection Selection favoring rapid population increase. r-Selection generally favors the production of large numbers of offspring with relatively little investment in each. (*Contrast with* **K-Selection**)

Recombination (also **Sexual Recombination**) The combining of genes from male sperm and female egg during fertilization. The new genetic combinations produced in this manner generate an enormous amount of genetic diversity upon which natural selection may operate.

Releasers Simple stimuli that elicit behaviors having a large genetic component. For example, the sight of the belly of a male stickleback fish causes an aggressive response in another male.

Resource A feature of the environment that contributes to an organism's fitness.

Ritualization   The evolutionary modification of a behavior pattern to serve communication.

Scent Marking   The deposition of **pheromones,** typically on a tree, bush, or rock.

Sexual Bimaturism   The occurrence of differences between males and females in the age at which sexual maturation occurs. Typically, the more competitive sex matures later.

Sexual Dimorphism   The occurrence of distinct differences between male and female, beyond the primary sexual characteristics.

Sociobiology   The study of the biological bases of social behavior, employing evolution as the basic explanatory tool.

Sociogram   A term coined by E.O. Wilson referring to a detailed cataloging of all the patterns of social behavior of a species.

Species   A population or series of populations of closely related organisms that are capable of exchanging genes under natural conditions.

Stabilizing Selection   Selection against the extremes, thereby maintaining the population around the mean.

Statistical Distribution of a Trait   The pattern revealed by examining the numbers of individuals in a population that possess a particular trait or that possess a trait to a particular extent. Evolutionary change is a change in the statistical distribution of genes in a population; it is apparent because of a change in the distribution of traits.

Synthetic Theory of Evolution   The most generally agreed upon interpretation of evolutionary change, based primarily on a synthesis of natural selection and genetics.

Territory   Any defended area. Alternatively, an area of exclusive occupancy.

Ultimate Causation.   The evolutionary factors conferring selective advantage on certain proximate mechanisms. Interpreting a behavior in terms of ultimate causation involves identifying the adaptive significance of that behavior.

References

Adler, A.
1932   *What Life Should Mean to You.* London: Allen & Unwin.

Alcock, J.
1975   *Animal Behavior: An Evolutionary Approach.* Sunderland: Sinnauer Associates.

Alexander, R. D.
1971   The search for an evolutionary philosophy of man. *Proceedings of the Royal Society of Victoria,* 84: 99–120.
1974   The evolution of social behavior. *Annual Review of Ecology and Systematics,* 5: 325–383.
1975   The search for a general theory of behavior. *Behavioral Science,* 20: 77–100.

Allee, W. C.
1938   *The Social Life of Animals.* New York: W. W. Norton.

Allen, L., et al.
1976   Sociobiology—Another biological determinism. *BioScience,* 26: 182–186.

Altmann, M.
1960   The role of juvenile elk and moose in the social dynamics of their species. *Zoologica,* 45: 35–39.

Altmann, S. A.
1974   Baboons, space, time and energy. *American Zoologist,* 14: 221–248

Altmann, S. A., and J. Altmann
1970   *Baboon Ecology: African Field Research.* Chicago: University of Chicago Press.

Anderson, W. W., and C. E. King
1970   Age-specific selection. *Proceedings of the National Academy of Sciences,* 66: 780–786.

Archer, J.
1970   Effects of population density on behaviour in rodents. In *Social Behaviour in Birds and Mammals* (J. Crook, ed.), pp. 169–210. New York: Academic.

Ardrey, R.
1966   *The Territorial Imperative.* New York: Atheneum.
1970   *The Social Contract.* New York: Atheneum.

Armitage, K. B.
1962   Social behaviour of a colony of the yellow-bellied marmot *(Marmota flaviventris). Animal Behaviour,* 10: 319–331.
1974   Male behaviour and territoriality in the yellow-bellied marmot. *Journal of Zoology,* 192: 233–265.

Armstrong, E. A.
1947   *Bird Display and Behaviour.* London: Lindsay Drummond.

Armstrong, J. T.
1965   Breeding home range in the nighthawk and other birds: its evolutionary and ecological significance. *Ecology,* 46: 610–629.

Barash, D. P.
1973a   The social biology of the Olympic marmot. *Animal Behaviour Monographs,* 6: 171–249.
1973b   Territorial and foraging behavior of pika *(Ochotona princeps)* in Montana. *American Midland Naturalist,* 12: 202–207.
1973c   Human ethology: Personal space reiterated. *Environment and Behavior,* 7: 67–72.
1974a   The evolution of marmot societies: A general theory. *Science,* 185: 415–420.
1974b   Social behavior of the hoary marmot *(Marmota caligata). Animal Behaviour,* 22: 257–262.
1974c   An adaptive advantage to winterflocking in the blackcapped chickadee, *Parus atricapillus. Ecology,* 55: 674–676.
1975a   Marmot alarm calling and the question of altruistic behavior. *American Midland Naturalist,* 94: 468–470.
1975b   Neighbor recognition in two "solitary" carnivores: The

raccoon *(Procyon lotor)* and the red fox *(Vulpes fulva). Science,* 185: 794–796.

1975c   Ecology of paternal behavior in the hoary marmot *(Marmota caligata)*: An evolutionary interpretation. *Journal of Mammalogy,* 56: 612–615.

1975d   Evolutionary aspects of parental behavior: The distraction behavior of the alpine accentor, *Prunella collaris. Wilson Bulletin,* 87: 367–373.

1975e   Behavioral individuality in the cichlid fish, *Tilapia mossambica. Behavioral Biology,* 13: 197–202.

1976a   The male response to apparent female adultery in the mountain bluebird, *Sialia currucoides:* An evolutionary interpretation. *American Naturalist,* in press.

1976b   Social behavior and individual differences in free-living Alpine marmots *(Marmota marmota). Animal Behavior,* 24: 27–35.

1976c   Some evolutionary aspects of parental behavior in animals and man. *American Journal of Psychology,* 89: 195-217.

1976d   What does sex really cost? *American Naturalist,* in press.

Barnett, S. A.
1958   An analysis of social behaviour in wild rats. *Proceedings of the Zoological Society of London,* 130: 107–152.

Bartholomew, G. A.
1952   Reproductive and social behavior of the northern elephant seal. *University of California Publications in Zoology,* 47: 369–472.

1970   A model for the evolution of pinniped polygyny. *Evolution,* 24: 546–559.

Bates, B. C.
1970   Territorial behavior in primates: A review of recent field studies. *Primates,* 11: 271–284.

Beach, F. A.
1950   The snark was a boojam. *American Psychologist,* 5: 115–124.

Bentley, D. R., and R. R. Hoy
1972   Genetic control of the neuronal network generating cricket *(Teleogryllus gryllus)* song patterns. *Animal Behaviour,* 20: 478–492.

Benzer, S.
1971   From the gene to behavior. *Journal of the American Medical Association,* 218: 1015–1026.
1973   Genetic dissection of behavior. *Scientific American,* 229: 24–37.

Bernstein, I. S.
1969   Spontaneous reorganization of a pigtail monkey group. *Proceedings of the Second International Congress of Primatology, Atlanta, Georgia,* 1: 48–51.

Blake, J.
1974   The changing status of women in developed countries. *Scientific American,* 231: 136–147.

Blurton-Jones, N.
1972   *Ethological Studies of Child Behaviour.* London: Cambridge University Press.

Bolles, R. C.
1970   Species-specific defense reactions and avoidance learning. *Psychological Review,* 77: 32–48.

Boorman, S. A., and P. R. Levitt
1972   Group selection on the boundary of a stable population. *Proceedings of the National Academy of Sciences,* 69: 2711–2713.
1973   Group selection on the boundary of a stable population. *Theoretical Population Biology,* 4: 85–128.

Bowlby, J.
1973   *Separation: Anxiety and Loss.* New York: Basic Books.

Bradbury, J.
1975   Social organization and communication. In *Biology of Bats* (W. Wimsatt, ed.). New York: Academic Press.

Breland, K., and M. Breland
1961   The misbehavior of organisms. *American Psychologist,* 16: 681–684.

Bronson, F. H.
1963   Some correlates of interaction rate in natural populations of woodchucks. *Ecology,* 44: 637–643.

Brown, J. L.
1963   Aggressiveness, dominance and social organization in the Stellar's jay. *Condor,* 65: 460–484.

| 1964 | The evolution of diversity in avian territorial systems. *Wilson Bulletin,* 76: 160–169. |

1964   The evolution of diversity in avian territorial systems. *Wilson Bulletin,* 76: 160–169.

1969   Territorial behavior and population regulation in birds: A review and re-evaluation. *Wilson Bulletin,* 81: 293–329.

1974   Alternate routes to sociality in jays—with a theory for the evolution of altruism and communal breeding. *American Zoologist,* 14: 63–80.

1975   *The Evolution of Behavior.* New York: W. W. Norton.

Brown, W. L., and C. O. Wilson

1956   Character displacement. *Systematic Zoology,* 5: 49–64.

Buechner, H. K., and H. D. Roth

1974   The lek system in Uganda kob antelope. *American Zoologist,* 14: 145–162.

Burt, W. H.

1943   Territoriality and home range concepts as applied to mammals. *Journal of Mammalogy,* 24: 346–352.

Burton, M.

1954   *Animal Courtship.* New York: Praeger.

Calhoun, J. B.

1962   Population density and social pathology. *Scientific American,* 206: 139–148.

Campbell, D. T.

1975   On the conflicts between biological and social evolution and between psychology and moral tradition. *American Psychologist,* 30: 1103–1126.

Carey, M., and V. Nolan

1975   Polygyny in indigo buntings: An hypothesis tested. *Science,* 190: 1296–1297.

Carl, E. A.

1971   Population control in arctic ground squirrels. *Ecology,* 52: 395–413.

Carpenter, C. R.

1934   A field study of the behavior and social relations of howling monkeys. *Comparative Psychology Monographs,* 10: 1–168.

1940   A field study in Siam of the behavior and social relations of the gibbon *(Hylobates lar). Comparative Psychology Monographs,* 16: 1–212.

1958   Territoriality: A review of concepts and problems. In *Behavior and Evolution* (A. Roe and G. Simpson, eds.), pp. 224–250. New Haven: Yale University Press.

Carrick, R.
1963   Ecological significance of territory size in the Australian magpie, *Gymnorhina tibiten. Proceedings of the International Ornithological Congress,* 13: 740–753.

Caughley, G.
1964   Social organization and daily activity of the red kangaroo and the green kangaroo. *Journal of Mammalogy,* 45: 429–436.

Chance, M. R. A.
1962   An interpretation of some agonistic postures: the role of "cut-off" acts and postures. *Symposium of Zoological Society of London,* 8: 71–89.

Chance, M. R. A., and A. P. Mead
1953   Social behavior and primate evolution. *Symposium of the Society for Experimental Biology,* 7: 395–439.

Charnov, E. L.
1976   Optimal foraging: Attack strategy of a mantid. *American Naturalist,* 110: 141–151.

Charnov, E. L., and J. R. Krebs
1974   The evolution of alarm calls: Altruism or manipulation? *American Naturalist,* 109: 107–112.

Chivers, D. J.
1972   The siamang and the gibbon in the Malay Peninsula. *Gibbon and Siamang,* 1: 103–135.

Chomsky, N.
1972   *Language and Mind.* New York: Harcourt, Brace and Jovanovich.

Christian, J. J.
1968   Endocrine-behavioral negative feed-back responses to increased population density. In *L'effet de Groupe Chez les Animaux* (R. Chauvin and C. Noirot, eds.), pp. 289–322. Paris: Centre National de la Recherche Scientifique.
1970   Social subordination, population density, and mammalian evolution. *Science,* 168: 84–90.

Christian, J. J., and D. E. Davis
1964   Endocrines, behavior and population. *Science,* 146: 1550–1560.

Clark, E., L. R. Aronson, and M. Gordon
1954   Mating behavior patterns in two sympatric species of xiphophorin fishes: Their inheritance and significance in sexual isolation. *Bulletin of the American Museum of Natural History,* 103: 135–226.

Clutton-Brock, T. H.
1974   Primate social organization and ecology. *Nature,* 250: 539–542.

Cody, M. L.
1969   Convergent characteristics in sympatric species: A possible relation to interspecific competition and aggression. *Condor,* 71: 222–239.
1971   Finch flocks in the Mohave Desert. *Theoretical Population Biology,* 2: 142–158.
1974   Optimization in ecology. *Science,* 183: 1156–1164.

Cole, L. C.
1954   The population consequences of life history phenomena. *Quarterly Review of Biology,* 29: 103–137.

Collias, N. E.
1943   Statistical analysis of factors which make for success in initial encounters between hens. *American Naturalist,* 77: 519–538.

Conder, P. J.
1949   Individual distance. *Ibis,* 91: 649–655.

Conner, R. L.
1972   Hormones, biogenic amines and aggression. In *Hormones and Behavior* (S. Levine, ed.), pp. 209–233. New York: Academic.

Constantz, G. D.
1975   Behavioral ecology of mating in the male Gila topminnow, *Poeciliopsis occidentalis. Ecology* 56: 966–973.

Cook, W. T., P. Siegel, and K. Hinkelmann
1972   Genetic analyses of male mating behavior in chickens: 2. Crosses among selected and control lines. *Behavior Genetics,* 2: 289–300.

Coulson, J. C.

1966   The influence of the pair-bond and age on the breeding biology of the kittiwake gull, *Rissa tridactyla. Journal of Animal Ecology,* 35: 269–279.

Crane, J.

1943   Display, breeding and relationship of the fiddler crabs. *Zoologica,* 28: 217–223.

Crook, J. H.

1965   The adaptive significance of avian social organizations. *Symposia of the Zoological Society of London,* 14: 181–218.

1970a   Social organization and the environment. *Animal Behaviour,* 18: 197–209.

1970b   The socio-ecology of primates. In *Social Behaviour in Birds and Mammals* (J. Crook, ed.). New York: Academic.

Crook, J. H., and J. S. Gartlan

1966   Evolution of primate societies. *Nature,* 210: 1200–1203.

Crow, J. F., and M. Kimura

1970   *An Introduction to Population Genetics Theory.* New York: Harper & Row.

Callen, E.

1957   Adaptations in the kittiwake to cliff-nesting. *Ibis,* 99: 275–302.

Darling, F. F.

1937   *A Herd of Red Deer.* London: Oxford University Press.

1938   *Bird Flocks and the Breeding Cycle: A Contribution to the Study of Avian Sociality.* Cambridge: Cambridge University Press.

Darwin, C.

1871   *The Descent of Man, and Selection in Relation to Sex.* New York: Appleton.

Davis, D. E., and F. Peek

1972   Stability of a population of male red-winged blackbirds. *Wilson Bulletin,* 84: 349–350.

Dawkins, R., and R.T. Carlisle

1976   Parental investment, mate desertion and a fallacy. *Nature,* 262: 131–132.

Deag, J. M., and J. H. Crook

1971   Social behaviour and "agonistic buffering" in the wild barbary macaque, *Macaca sylvang. Folia Primatologica,* 15: 183–200.

DeFries, J. C., and G. E. McClearn
1970   Social dominance and Darwinian fitness in the laboratory mouse. *American Naturalist,* 104: 408–411.

DeVore, I., and K. R. L. Hall
1965   Baboon ecology. In *Primate Behavior* (I. DeVore, ed.), pp. 20–52. New York: Holt, Rinehard and Winston.

Dilger, W.
1962   The behavior of lovebirds. *Scientific American,* 206: 88–98.

Dobzhausky, T.
1951   *Genetics and the Origin of Species.* New York: Columbia University Press.
1970   *Genetics of the Evolutionary Process.* New York: Columbia University Press.
1976   The myths of genetic predestination and of tabula rasa. *Perspectives in Biology and Medicine,* 156–170.

Dolland, J., N. E. Miller, O. H. Mowrer, G. H. Sears, and R. R. Sears
1939   *Frustration and Aggression.* New Haven: Yale University Press.

Downhower, J. F., and K. B. Armitage
1971   The yellow-bellied marmot and the evolution of polygamy. *American Naturalist,* 105: 355–370.

Dunbar, R. I. M., and E. P. Dunbar
1976   Contrasts in social structure among black-and-white colobus monkey groups. *Animal Behaviour,* 24: 84–92.

Eaton, T. H.
1970   *Evolution.* New York: Norton.

Eberhard, W. G.
1972   Altruistic behavior in a sphecid wasp: Support for kin-selection theory. *Science,* 172: 1390–1391.

Ehrenkranz, J., E. Bliss, and M. Sheard
1974   Plasma testosterone: correlation with aggressive behavior and social dominance in man. *Psychosomatic Medicine,* 36: 469–475.

Ehrlich, P. R., and R. W. Holm
1963   *The Process of Evolution.* New York: McGraw-Hill.

Ehrman, L., and P. A. Parsons
1976   *The Genetics of Behavior.* Sunderland, Mass.: Sinnauer Associates.

Eibl-Eibesfeldt, I.
1961   The fighting behavior of animals. *Scientific American,* 205: 112–121.
1966   Das Verteidigen der Eiablageplatze bei der Hood-Meerechse. *Zeitschrift für Tierpsychologie,* 23: 627–631.
1975   *Ethology, the Biology of Behavior.* New York: Holt, Rinehard and Winston.

Eiseley, L.
1958   *Darwin's Century.* New York: Doubleday.

Eisenberg, J. F.
1966   The social organization of mammals. *Handbuch der Zoologie,* 10: 1–92.
1967   A comparative study in rodent ethology with emphasis on evolution of social behavior. *Proceedings of the United States National Museum, Washington, D. C.,* 122: 1–51.

Eisenberg, J. F., and R.E. Kuehn
1966   The behavior of *Ateles geoffroyi* and related species. *Smithsonian Miscellaneous Collections,* 151: 1–63.

Eisenberg, J. F., N. A. Muckenhirn, and R. Rudran
1972   The relation between ecology and social structure in primates. *Science,* 176: 863–874.

Emlen, J. M.
1966   Natural selection and human behavior. *Journal of Theoretical Biology,* 12: 410–418.
1972   *Ecology: An Evolutionary Approach.* Reading: Addison-Wesley.

Emlen, S. T., and N. J. Demong
1975   Adaptive significance of synchronized breeding in a colonial bird: A new hypothesis. *Science,* 188: 1029–1031.

Erickson, C. J., and P. G. Zenone
1976   Courtship differences in male ring doves: Avoidance of cuckoldry? *Science,* 192: 1353–1354.

Errington, P. L.
1963   *Muskrat Populations.* Ames, Iowa: Iowa State University Press.

Estes, R. D.

1966 Behaviour and life history of the wildebeest. *Nature,* 212: 999–1000.

1974 Social organization of the African Bovidae. In *The Behaviour of ungulates and its Relation to Management* (V. Geist and F. Walther, eds.), pp. 166–205. Morges: I.U.C.N. Publications.

Estes, R. D., and J. Goddard

1967 Prey selection and hunting behavior of the African wild dog. *Journal of Wildlife Management,* 31: 52–70.

Etkin, W., ed.

1964 *Social Behavior and Organization Among Vertebrates.* Chicago: University of Chicago Press.

Evans, H. E.

1962 The evolution of prey-carrying mechanisms in wasps. *Evolution,* 16: 468–483.

Ewer, R. F.

1968 *Ethology of Mammals.* London: Logos Press.

Fagen, R. M.

1972 An optimal life-history strategy in which reproductive effort decreases with age. *American Naturalist,* 106: 258–261.

Farner, D. S., and J. R. King, eds.

1971 *Avian Biology.* New York: Academic.

Fiedler, K.

1954 Vergleichende Verhaltensstudien an Seenadeln, Schlang enn adeln und Seepferd chen (Syngnathidge). *Zeitschrift für Tierpsychologie,* 11: 358–416.

Fisher, J., and R. A. Hinde

1948 The opening of milk bottles by birds. *British Birds,* 42: 347–357.

Fisher, R. A.

1930 *The Genetical Theory of Natural Selection.* Oxford: Clarendon.

Fitch, H. S., and H. W. Shirer

1970 A radiotelemetric study of spatial relationships in the oppossum. *American Midland Naturalist,* 84: 170–186.

Fox, M. W.
1970   A comparative study of the development of facial expressions in canids: Wolf, coyote and foxes. *Behaviour,* 36: 49–73.

Fox, R.
1972   Alliance and constraint: Sexual selection in the evolution of human kinship systems. In *Sexual Selection and the Descent of Man* (B. Campbell, ed.). Chicago: Aldine.

Fraenkel, G. S., and D. L. Gunn
1940   *The Orientation of Animals.* London: Oxford University.

Fretwell, S.
1969   Dominance behavior and winter habitat distribution in juncos *(Junco hyemalis). Bird-banding,* 40: 1–25.

Fretwell, S. D., and H. L. Lucas
1969   On territorial behavior and other factors influencing habitat distribution in birds. *Acta biotheoretica,* 19: 16–36.

Freud, S.
1920   *Beyond the Pleasure Principle.* London: Hogarth Press.

Fromm, E.
1973   *The Anatomy of Human Destructiveness.* New York: Holt, Rinehart and Winston.

Fry, C. H.
1972   The social organization of bee-eaters (Meropidoe) and cooperative breeding in hot-climate birds. *Ibis,* 114.

Gadgil, M.
1975   Evolution of social behavior ·through interpopulation selection. *Proceedings of the National Academy of Sciences,* 72: 1199–1201.

Gadgil, M., and W. H. Bossert
1970   Life history consequences of natural selection. *American Naturalist,* 104: 1–24.

Galton, F.
1871   Gregariousness in cattle and men. *Macmillan's Magazine,* 23: 353.

Garcia, J., W. G. Hankins, and K. W. Rusiniak
1974   Behavioral regulation of the milieu interne in man and rat. *Science,* 185: 824–831.

Gardner, R. A., and B. T. Gardner
1971   Two-way communication with an infant chimpanzee. In

Behavior of Non-Human Primates (A. Schrier and F. Stollnitz, eds.), pp. 117–184. New York: Academic.

Gartlan, J. S.
1968   Structure and function in primate society. *Folia Primatologica,* 8: 89–120.

Geist, V.
1971   *Mountain Sheep: A Study in Behavior and Evolution.* Chicago: University of Chicago Press.
1974   On the relationship of social evolution and ecology in ungulates. *American Zoologist,* 14: 205–220;

Gilliard, E. T.
1969   *Birds of Paradise and Bower Birds.* Garden City, N. Y.: Natural History Press.

Glas, P.
1960   Factors governing density in the claffinch *(Fringilla coelebs)* in different types of wood. *Archives Neerlandaises de Zoologie,* 13: 466–472.

Goss-Custard, J. D.
1970   Feeding dispersion in some overwintering wading birds. In *Social Behaviour in Birds and Mammals* (J. H. Crook, ed.), pp. 3–35. New York: Academic.

Grant, V.
1963   *The Origin of Adaptations.* New York: Columbia.

Greenberg, B.
1947   Some relations between territory, social hierarchy, and leadership in the green sunfish *(Lepomis cyanellus).* *Physiological Zoology,* 20: 267–299.

Greene, P. J., and D. P. Barash
1976   Genetic basis of behavior—especially of altruism. *American Psychologist,* 31: 359–361.

Guhl, A. M., N. E. Collias, and W. C. Allee
1945   Mating behavior and the social hierarchy in small flocks of white leghorns. *Physiological Zoology,* 18: 365–390.

Guhl, A. M., and G. J. Fisher
1969   *The Behavior of Domestic Animals* (E. Hafez, ed.), pp. 513–553. Baltimore: Williams and Wilkens.

Guthrie, E. R.
1935   *The Psychology of Learning.* New York: Harper & Row.

Hailman, J. P.
1964   Breeding synchrony in the equatorial swallow-tailed gull. *American Naturalist,* 98: 79–83.

Haldane, J. B. S.
1932   *The Causes of Evolution.* London: Longmans, Green.

Hall, E. T.
1966   *The Hidden Dimension.* Garden City, N. Y.: Doubleday.

Hamilton, W. D.
1964   The genetical theory of social behaviour: I. and II. *Journal of Theoretical Biology,* 7: 1–52.
1967   Extraordinary sex ratios. *Science,* 156: 477–488.
1971   Geometry for the selfish herd. *Journal of Theoretical Biology,* 31: 295–311.
1970   Selfish and spiteful behaviour in an evolutionary model. *Nature,* 228: 1218–1220.
1975   Innate social aptitudes of man: An approach from evolutionary genetics. In *Biosocial Anthropology* (R. Fox, ed.), pp. 133–155. New York: Wiley.

Hamilton, W. J.
1973   *Life's Color Code.* New York: McGraw-Hill.

Hardin, G.
1968   The tragedy of the commons. *Science,* 162: 1243–1248.

Harlow, H. F., and M. K. Harlow
1962   Social deprivation in monkeys. *Scientific American,* 207: 136–146.

Healey, M. C.
1967   Aggression and self-regulation of population size in deermice. *Ecology,* 48: 377–392.

Hediger, H.
1955   *Studies of the Psychology and Behavior of Captive Animals in Zoos and Circuses.* New York: Criterion Books.

Heinroth, O.
1910   Beitrage zur Biologie, namentlich Ethologie und Physiologie der Anatiden. *International Ornithologisches Kongress,* 5: 589–702.

Heller, H. C.
1971   Altitudinal zonation of chipmunks *(Eutamias)*: Interspecific aggression. *Ecology,* 52: 312–329.

Hensley, M. M., and J. B. Cope
1951   Further data on removal and repopulation of the breeding birds in a spruce fir community. *Auk*, 68: 483–493.

Hilden, O., and S. Vuolanto
1972   Breeding biology of the red-necked phalarope *Phalaropus lobatus* in Finland. *Ornis Fennicg*, 49: 57–85.

Hinde, R. A.
1956   The biological significance of the territories of birds. *Ibis*, 98: 340–369.
1970   *Animal Behaviour: A Synthesis of Ethology and Comparative Psychology*. New York: McGraw-Hill.
1972   *Non-Verbal Communication*. Cambridge: Cambridge University Press.

Hinde, R. A., and L. M. Davies
1972   Removing infant rhesus from mother for 13 days compared with removing mother from infant. *Journal of Child Psychology and Psychiatry*, 13: 227–237.

Hinde, R. A., and Y. Spencer-Booth
1971   Effects of brief separation from mother on rhesus monkeys. *Science*, 173: 111–118.

Hinde, R. A., and J. Stevenson-Hinde
1973   *Constraints on Learning*. New York: Academic.

Hirsch, J.
1963   Behavior genetics and individuality understood. *Science*, 142: 1436–1442.

Hirsch, J., and L. Erlenmeyer-Kimling
1962   Individual differences in behavior and their genetic basis. In *Roots of Behavior* (E. L. Bliss, ed.), pp. 3–23. New York: Harper & Row.

Horn, H. S.
1968   The adaptive significance of colonial nesting in the Brewer's blackbird *(Euphagus cyanocephalus)*. *Ecology*, 49: 682–694.

Howard, H. E.
1920   *Territory in Bird Life*. London: John Murray.

Hrdy, S. B.
1974   Male-male competition and infanticide among the langurs *(Presbytis entellus)* of Abu. Rajasthan. *Folia Primatologica*, 22: 19–58.

Hrdy, S.B., and D.B. Hrdy
1976   Hierarchical relations among female Hanuman langurs (Primates: Colobinae, *Presbytis entellus*). *Science,* 193: 913–915.

Hull, C. L.
1943   *Principles of Behavior.* New York: Appleton-Century-Crofts.

Humphries, D. A., and P. M. Driver
1967   Erratic display as a device against predators. *Science,* 156: 1767–1768.

Huxley, J. S.
1923   Courtship activities in the red-throated diver *(Colymbus stellatus);* together with a discussion of the evolution of courtship in birds. *Journal of the Linnaen Society of London, Zoology,* 35: 253–292.
1934   A natural experiment on the territorial instinct. *British Birds,* 27: 270–277.
1938   The present standing of the theory of sexual selection. In *Evolution: Essays on Aspects of Evolutionary Biology Presented to Professor E. S. Goodrich on his Seventieth Birthday* (G. deBeeried.), pp. 11–42. Oxford: Clarendon.
1942   *Evolution: The Modern Synthesis.* London: Allen & Unwin.

Jarman, P. J.
1974   The social organization of antelope in relation to their ecology. *Behavior,* 58: 215–267.

Jay, P. C.
1968   *Primates: Studies in Adaptation and Variability.* New York: Holt, Rinehard and Winston.

Jenni, D. A.
1974   Evolution of polyandry in birds. *American Zoologist,* 14: 129–144.

Johnsgard, P. A.
1967   Dawn rendezvous on the lek. *Natural History,* 76: 16–21.

Jolly, A.
1972   *The Evolution of Primate Behavior.* New York: Macmillan.

Kalleberg, H.
1950   Observations in a stream tank of territoriality and competition in juvenile salmon and trout. *Reports of the Institute of Freshwater Research, Drottingholm,* 39: 55–98.

Katz, P. L.
1974   A long-term approach to foraging optimization. *American Naturalist,* 108: 758–782.

Kaufmann, J. H.
1962   Ecology and social behavior of the coati, *Nasua narica,* on Barro Colorado Island, Panama. *University of California Publications in Zoology,* 60: 95–222.

Kawai, M.
1965   Newly acquired pre-cultural behavior of the natural troop of Japanese monkeys on Koshima Islet. *Primates,* 6: 1–30.

Kawamura, S.
1967   Aggression as studied in troops of Japanese monkeys. In *Brain Function, Vol. 5, Aggression and Defense, Neural Mechanisms and Social Patterns* (C. Clemente and D. Lindsley, eds.), pp. 195–223. Berkeley: University of California Press.

Kessel, E. L.
1955   The mating activities of balloon flies. *Systematic Zoology,* 4: 97–104.

Kettlewell, H. B. D.
1956   Further selection experiments on industrial melanism in the Lepidoptera. *Heredity,* 10: 287–301.

King, J. A.
1955   Social behavior, Social Organization, and Population Dynamics in Black-Tailed Prairiedog Town in the Black Hills of South Dakota. Contributions from the Laboratory of Vertebrate Biology, University of Michigan, no. 67, Ann Arbor.

Kluyver, H. M., and L. Tinbergen
1953   Territory and regulation of density in titmice. *Archives Neerlandaises de Zoologie,* 10: 265–289.

Köhler, W.
1925   *The Mentality of Apes.* New York: Harcourt, Brace.
1947   *Gestalt Psychology.* New York: Liveright Publishing.

Kortland, A.
1972   New perspectives on ape and human evolution. *Stichung voor Psychobiologie, Univ. van Amsterdam,* 100 pp.

Kramer, G.

1946   Veranderungen von Nachkommenziffer und Nachkommengrosse sowie der Altersuerteilung von Inseleidechsen. *Zeitschrift für Naturforschen,* 1: 700–710.

Krebs, J. R.

1971   Territory and breeding density in the great tit, *Parus major. Ecology,* 52: 2–22.

Kretchmer, N.

1972   Lactose and lactase. *Scientific American*, 227:70–78.

Kruuk, H.

1964   Predators and anti-predator behaviour of the black-headed gull *(Larus ridibandus). Behaviour* supplement, 11: 1–129.

1972   *The Spotted Hyena.* Chicago: University of Chicago Press.

Kuhn, T.S.

1962   *The Structure of Scientific Revolutions.* Chicago: University of Chicago Press.

Kummer, H.

1968   *Social Organization of Hamadrayas Baboons: A Field Study.* Chicago: University of Chicago Press.

1971   *Primate Societies: Group Techniques of Ecological Adaptation.* Chicago: Aldine-Atherton.

Lack, D.

1947   *Darwin's Finches.* New York: Cambridge University Press.

1968   *Ecological Adaptations for Breeding in Birds.* London: Methuen.

Lagerspetz, K.

1964   *Studies on the Aggressive Behaviour of Mice.* Helsinki: Suomalainen Tiedeakatemia.

Lancaster, J. B.

1971   Play-mothering: the relations between juvenile females and young infants among free-ranging vervet monkeys *(Cercopithecus aethiops). Folia Primatologica,* 15: 161–182.

LeBouef, B. J.

1974   Male-male competition and reproductive success in elephant seals. *American Zoologist,* 14: 163–176.

Lee, R. B., and I. DeVore, eds.

1968   *Man the Hunter.* Chicago: Aldine.

Lehrman, D. S.

1953   A critique of Konrad Lorenz' theory of instinctive behavior. *Quarterly Review of Biology,* 28: 337–363.

1970   Semantics and conceptual issues in the nature-nurture problem. In *Development and Evolution of Behavior* (L. R. Aronson, et al., eds.). San Francisco: W. H. Freeman.

Leigh, E. G.

1970   Sex ratio and differential mortality between the sexes. *American Naturalist,* 104: 205–210.

Lemmetyinen, R.

1971   Nest defence behaviour of Common and Arctic Terns and its effects on the success achieved by predators. *Ornis Fennica,* 48: 12–24.

Leuthold, W.

1966   Variations in territorial behavior of Uganda Kob. *Behaviour,* 27: 215–258.

Levins, R.

1970   Extinction. In *Some Mathematical Questions in Biology* (M. Gerstenhaber, ed.), pp. 77–107. Am. Mathematical Society.

Levi-Strauss, C.

1969   *The Elementary Structures of Kinship* (R. Needham, ed.). Boston: Beacon Press.

Lewontin, R. C.

1970   The units of selection. *Annual Review of Ecology and Systematics,* 1: 1–18.

1974   *The Genetic Basis of Evolutionary Change.* New York: Columbia University Press.

Li, C.C.

1955   *Population Genetics.* Chicago: University of Chicago Press.

Lidicker, W. Z.

1962   Emigration as a possible mechanism permitting the regulation of population density below carrying capacity. *American Naturalist,* 96: 29–33.

1965   Comparative study of density regulation in confined populations of four species of rodents. *Researches on Population Ecology,* 7: 57–72.

Lin, N., and C. D. Michener

1972   Evolution of sociality in insects. *Quarterly Review of Biology,* 47: 131–159.

Lindzey, G., J. Loehlin, M. Manosevitz, and D. Thiessen
1971   Behavioral genetics. *Annual Review of Psychology*, 22: 39–94.

Lindzey, G., and D. Thiessen
1970   *Contributions to Behavior-Genetic Analysis: The Mouse as a Prototype.* New York: Appleton.

Lockard, R. B.
1971   Reflections on the fall of comparative psychology: Is there a message for us all? *American Psychologist*, 26: 168–179.

Loeb, J.
1906   *The Dynamics of Living Matter.* New York: Columbia University Press.

Lorenz, K. Z.
1950   The comparative method of studying innate behaviour patterns. In *Psychological Mechanisms in Animal Behaviour*, pp. 221–268. New York: Academic.
1952   *King Solomon's Ring.* London: Methuen.
1958   The evolution of behavior. *Scientific American*, 199: 67–68.
1963   *On Aggression.* New York: Harcount, Brace and World.
1974   Analogy as a source of knowledge. *Science*, 185: 229–233.

MacArthur, R. H.
1965   Ecological consequences of natural selection. In *Theoretical and Mathematical Biology* (T. Waterman and H. Morowitz, eds.), pp. 388–397. New York: Blaisdell.
1972   *Geographical Ecology: Patterns in the Distribution of Species.* New York: Harper and Row.

MacKinnon, J.
1974   The behaviour and ecology of wild orang-utans *(Pongo pygmaeus)*. *Animal Behaviour*, 22: 3–74.

Mackintosh, J. H.
1970   Territory Formation by laboratory mice. *Animal Behaviour*, 18: 177–183.

Manning, A.
1965   Drosophila and the evolution of behavior. In *Viewpoints in Biology* (J. Carthy and C. Duddington, eds.), pp. 125–169. London: Butterworth.

Marler, P. R.
1956   Behaviour of the chaffinch, *Frivgilla coelebs. Behaviour* supplement, 5: 1–184.

Martin, S. G.
1974   Adaptations for polygynous breeding in the bobolink, *Dolinchonyx oryzivorus. American Zoologist,* 14: 109–120.

Mather, K., and B. J. Harrison
1949   The manifold effect of selection. *Heredity,* 3: 1–52, 131ff.

Maynard Smith, J.
1964   Group selection and kin selection. *Nature,* 201: 1145–1147.
1966   *The Theory of Evolution.* Baltimore: Penguin.
1971   What use is sex? *Journal of Theoretical Biology,* 30: 319ff.
1974   The theory of games and the evolution of animal conflicts. *Journal of Theoretical Biology,* 47: 209–221.

Maynard Smith, J., and G. A. Parker
1976   The logic of asymmetric contests. *Animal Behaviour,* 24.

Maynard Smith, J., and G. R. Price
1973   The logic of animal conflict. *Nature,* 246: 15–18.

Maynard Smith, J., and M. G. Ridpath
1972   Wife sharing in the Tasmanian native hen, *Tribonyx mortierii:* A case of kin selection? *American Naturalist,* 106.

Mayr, E.
1963   *Animal Species and Evolution.* Cambridge: Harvard University Press.
1970   *Populations, Species, and Evolution.* Cambridge, Mass.: Harvard University Press.

McClearn, G. E.
1963   The inheritance of behavior. In *Psychology in the Making* (L. J. Postman, ed.), pp. 144–252. New York: Knopf.

McGrew, W. C.
1972   *An Ethological Study of Children's Behavior.* London: Academic.

McKay, F. E.
1971   Behavioral aspects of population dynamics in unisexual-bisexual *Poeciliopsis Ecology,* 52: 778–790.

McNab, B. K.

1963   Bioenergetics and the determination of home range size. *American Naturalist,* 97: 133–140.

Mech, L. D.

1970   *The Wolf: The Ecology and Behavior of an Endangered Species.* Garden City, N. Y.: Natural History Press.

Meyerriecks, A. J.

1960   Comparative breeding behavior of four species of North American herons. Publ. #2, The Nuttall Ornithological Club, Cambridge, MA. 158 pages.

Michael, R. P., and J. H. Crook, eds.

1973   *Comparative Ecology and Behaviour of Primates.* New York: Academic.

Michener, C. D.

1974   *The Social Behavior of the Bees: A Comparative Study.* Cambridge: Harvard University Press.

Michener, C. D., and D. J. Brothers

1974   Were workers of eusocial Hymenoptera initially altruistic or oppressed? *Proceedings of the National Academy of Sciences,* 71: 671–674.

Michener, G. R.

1973   Field observations on the social relationships between adult female and juvenile Richardson's ground squirrels. *Canadian Journal of Zoology,* 51: 33–38.

Milgram, S.

1974   *Obedience to Authority: An Experimental View.* New York: Harper & Row.

Mohr, H.

1960   Zum Erkenner von Raubvogeln, inbesondere von Sperber und Baum Falk, durch Kleinnvogeln. *Zeitschrift für Tierpsychologie,* 17: 686–699.

Moltz, H., ed.

1971   *The Ontogeny of Vertebrate Behavior.* New York: Academic.

Moment, G.

1962   Reflexive selection: a possible answer to an old puzzle. *Science,* 136: 262–263.

Money, J., and Eberhandt, A. A.
1972   *Man and Woman, Boy and Girl: The Differentiation and Dimorphism of Gender Identity from Conception to Maturity.* Baltimore: Johns Hopkins University Press.

Montagu, M. F. A., ed.
1968   *Man and Aggression.* Oxford: Oxford University Press.

Moody, P. A.
1970   *Introduction to Evolution.* New York: Harper and Row.

Morse, D. H.
1970   Ecological aspects of some mixed species foraging flocks of birds. *Ecological Monographs,* 40: 119–168.
1974   Niche breadth as a function of social dominance. *American Naturalist,* 108: 818–830.

Murray, B. G.
1971   The ecological consequences of interspecific territorial behavior in birds. *Ecology,* 52: 414–423.

Murton, R. K., A. J. I. Saacson, and N. J. Westwood
1966   The relationships between woodpigeons and their clover food supply and the mechanism of population control. *Journal of Applied Ecology,* 3: 55–96.

Nice, M. M.
1937   Studies in the life history of the song sparrow: Part I, A population study of the song sparrow. *Transactions of the Linnean Society of New York,* 4: 1–247.
1943   Studies in the life history of the song sparrow. *Transactions of the Linnean Society of New York,* 6: 1–328.

Noble, G. K.
1936   Courtship and sexual selection of the flicker *(Colaptes aurarus Luteus). Auk,* 53: 269–282.

Orians, G. H.
1961   The ecology of blackbird (Agelaius) social systems. *Ecological Monographs,* 31: 285–312.
1969   On the evolution of mating systems in birds and mammals. *American Naturalist,* 103: 589–603.

Orians, G. H., and M. F. Willson
1964   Interspecific territories of birds. *Ecology,* 45: 736–745.

Otte, D.
1974   Effects and functions in the evolution of signalling systems. *Annual Review of Ecology and Systematics,* 5: 385ff.

355

Parker, G. A.
1974   Assessment strategy and the evolution of animal conflicts. *Journal of Theoretical Biology*, 47: 223–243.

Parker, G. A., R. R. Baker, and V. G. F. Smith
1972   The origin and evolution of gamete dimorphism and the male-female phenomenon. *Journal of Theoretical Biology*, 36: 529–553.

Peek, F. W.
1971   Seasonal change in the breeding behavior of the male red-winged blackbird. *Wilson Bulletin*, 83: 383–395.

Perrins, C. M.
1965   Population fluctuations and clutch size in the great tit. *Parus major. Journal of Animal Ecology*, 34: 601–647.

Petit, C., and L. Ehrman
1969   Sexual selection in *Drosophila. Evolutionary Biology*, 3: 177–223.

Piaget, J.
1973   *The Child and Reality: Problems of Genetic Psychology.* New York: Grossman.

Pianka, E. R.
1970   On *r*-and *K*-selection. *American Naturalist*, 104: 592–297.
1974   *Evolutionary Ecology.* New York: Harper & Row.

Pitelka, F. A.
1959   Numbers, breeding schedule, and territoriality in pectoral sandpipers of northern Alaska. *Condor*, 61: 233–264.

Pitelka, F., R. Holmes, and S. Maclean
1974   Ecology and evolution of social organization in arctic sandpipers. *American Zoologist*, 14: 185–204.

Power, H. W.
1975   Mountain bluebirds: Experimental evidence against altruism. *Science*, 189: 142–143.

Premack, D.
1971   Language in the chimpanzee. *Science*, 172: 808–822.

Ralph, C. J., and C. A. Pearson
1971   Correlation of age, size of territory, plumage, and breeding success in white-crowned sparrows. *Condor*, 73: 77–80.

Ricklefs, R. E.
1973   *Ecology.* Newton, Mass.: Chiron Press.

Ripley, S. D.
1952   Territory and sexual behavior in the great Indian rhinoceros, a speculation. *Ecology,* 33: 570–573.
1961   Aggressive neglect as a factor in interspecific competition in birds. *Auk,* 78: 366–371.

Robertson, D. R.
1972   Social control of sex reversal in a coral-reef fish. *Science,* 177: 1007–1009.

Rosenblum, L. A.
1970   *Primate Behavior: Developments in Field and Laboratory Research.* New York: Academic.

Rothenbuhler, W.
1964   Behavior genetics of nest cleaning in honey bees: Responses of F1 and backcross generations to disease-killed brood. *American Zoologist,* 4: 111–123.

Rowell, T. E.
1967   Variability in the social organization of primates. In *Primate Ethology* (D. Morris, ed.). Chicago: Adine.

Rowell, T. E., R. A. Hinde, and Y. Spencer-Booth
1964   "Aunt"—infant interaction in captive rhesus monkeys. *Animal Behaviour,* 12: 219–226.

Rowley, I.
1965   The life history of the superb blue wren, *Malurus cyaneus. Emu,* 64: 251–297.

Sackett, G. P.
1966   Monkeys reared in isolation with pictures as visual input: evidence for an innate releasing mechanism. *Science,* 154: 1468–1473.

Sadleir, R. M. F. S.
1965   The relationship between agonistic behaviour and population changes in the deermouse, *Peromyscus maniculatus. Journal of Animal Ecology,* 34: 331–352.

Sahlins, M. D.
1965   On the sociology of primitive exchange. In *The Relevance of Models for Social Anthropology* (M. Banton, ed.), pp. 139–236. London: Tavistock.

Schaller, G. B.
1963   *The Mountain Gorilla: Ecology and Behavior.* Chicago: University of Chicago Press.
1972   *The Serengeti Lion: A Study of Predator-Prey Relations.* Chicago: University of Chicago Press.

Schaller, G. B., and G. R. Lowther
1969   The relevance of carnivore behavior to the study of early hominoids. *Southwestern Journal of Anthropology,* 25.

Schenkel, R.
1966   Zum Problem der Territorialitat und des Markierens dei Saugern—am Beispiel des Schwarzen Nashorns und des Lowens. *Zeitschrift für Tierpsychologie,* 23: 539/626.

Schoener, T. W.
1968   Sizes of feeding territories among birds. *Ecology,* 49: 123–141.
1971   Theory of feeding strategies. *Annual Review of Ecology and Systematics,* 2: 369–404.

Scott, J. P.
1958   *Aggression.* Chicago: University of Chicago Press.

Scott, J. P., and J. L. Fuller
1965   *Genetics and the Social Behavior of the Dog.* Chicago: University of Chicago Press.

Searle, L. V.
1949   The organization of hereditary maze-brightness and maze-dullness. *Genetic Psychology Monographs,* 39: 279 ff.

Sebeok, T. H., ed.
1968   *Animal Communication.* Bloomington: Indiana Univ. Press.

Seilacher, A.
1967   Fossil behavior. *Scientific American,* 217: 72–80.

Selander, R. K.
1965   On mating systems and sexual selection. *American Naturalist,* 99: 129–140.
1966   Sexual dimorphism and differential niche utilization in birds. *Condor,* 68: 113–151.
1970   Behavior and genetic variation in natural populations. *American Zoologist,* 10: 53–66.

Seligman, M. E. P., and J. L. Hager
1972   *Biological Boundaries of Learning.* New York: Appleton-Century-Crofts.

Selye, H.
1956　　*The Stress of Life.* New York: McGraw-Hill.

Sexton, O. J.
1960　　Some aspects of the behavior and of the territory of a dendrobatid frog, *Prostherapis Trinitatis. Ecology,* 41: 107–115.

Sharpe, R. S., and P. A. Johnsgard
1966　　Inheritance of behavioral characters in F2 mallard × pintail hybrids. *Behaviour,* 27: 259–272.

Sheppard, P. M.
1960　　*Natural Selection and Heredity.* New York: Harper and Row.

Sherrington, E. S.
1906　　*The Intergrative Action of the Nervous System.* New York: Cambridge University Press.

Shettleworth, S. J.
1972　　Constraints on learning. In *Advances in the Study of Behavior,* vol. 4 (D. S. Lehram et al., eds.), pp. 1–68. New York: Academic.

Siegel, P.
1972　　Genetic analysis of male mating behavior in chickens, 1. Artificial Selection. *Animal Behaviour,* 20: 564–570.

Simpson, G. G.
1949　　*The Meaning of Evolution.* New Haven: Yale University Press.
1953　　*The Major Features of Evolution.* New York: Columbia University Press.

Singh, S. D.
1969　　Urban monkeys. *Scientific American,* 221: 108–115.

Skinner, B. F.
1938　　*The Behavior of Organisms: An Experimental Analysis.* New York: Appleton-Century-Crofts.

Skutch, A. F.
1935　　Helpers at the nest. *Auk,* 52: 257–273.
1961　　Helpers among birds. *Condor* 63: 198–226.

Smith, A.
1975　　*Powers of Mind.* New York: Random House.

Smith, C. C.

1968   The adaptive nature of social organization in the genus of tree squirrels *Tamiasciurus*. *Ecological Monographs,* 38: 31–63.

Smith, N. G.

1968   The advantages of being parasitized. *Nature,* 219: 690–694.

Snow, D. W.

1963   The evolution of monakin displays. *Proceedings of the Thirteenth International Ornithological Congress,* pp. 533–561.

Southwick, C. H.

1969   Aggressive behaviour of rhesus monkeys in natural and captive groups. In *Aggressive Behaviour* (S. Garattini and E. Sigg, eds.). Amsterdam: Excerpta Medica.

Southwick, C. H., M. F. Siddiqi, M. Y. Farooqui, and B. C. Pal

1976   Effects of artificial feeding on aggressive behaviour of rhesus monkeys in India. *Animal Behaviour,* 24: 11–15.

Sparr, E.

1974   Individual differences in aggressiveness of adelie penguins. *Animal Behaviour,* 22: 611–616.

Spence, K. W.

1960   *Behavior Theory and Learning.* Englewood Cliffs: Prentice-Hall.

Spiess, E. B.

1970   Mating propensity and its genetic basis in *Drosophila.* In *Essays in Evolution and Genetics in Honor of Theogosius Dobzhansky* (M. Hecht and W. Stere, eds.), pp. 315–379. New York: Appleton-Century-Crofts.

Spieth, H. T.

1968   Evolutionary implications of sexual behavior in *Drosophila. Evolutionary Biology,* 2: 157–193.

Stebbins, G. L.

1971   *Processes of Organic Evolution.* Englewood Cliffs: Prentice-Hall.

Stefanski, R. A.

1967   Utilization of the breeding territory in the black-capped chickadee. *Condor,* 69: 259–267.

Stenger, J.
1958   Food habits and available food of ovenbirds in relation to territory size. *Auk,* 75: 335–346.

Sugiyama, Y.
1967   Social organization of hanuman langurs. In *Social Communication among Primates* (S. Altmann, ed.), pp. 221–236. Chicago: University of Chicago Press.

Suzuki, A.
1971   Carnivory and cannibalism observed among forest-living chimpanzees. *Journal of the Anthropological Society of Nippon,* 79: 30–48.

Taber, R. D., and R. F. Dassmann
1957   The dynamics of three natural populations of the deer, *Odoicoileus hemionus columbianus. Ecology,* 38: 233–246.

Tenaza, R.
1971   Behavior and nesting success relative to nest location in Adelie penguins *(Pygoscelis adeliae). Condor,* 73: 81–92.

Thornhill, R.
1976   Sexual selection and paternal investment in insects. *American Naturalist,* 110: 153–163.

Tiger, L., and R. Fox
1966   The zoological perspective in social science. *Man,* 1: 75–81.

Tiger, L., and J. Shepher
1975   *Women in the Kibbutz.* New York: Harcount, Brace, Jonanovich.

Tinbergen, N.
1963   The shell menace. *Natural History,* 72: 28–35.
1968   On war and peace in animals and man. *Science,* 160: 1411–1418.

Tinkle, D. W.
1969   The concept of reproductive effort and its relation to the evolution of life histories of lizards. *American Naturalist,* 103: 501–516.

Tokuda, K., and G. D. Jensen
1968   The leader's role in controlling aggressive behavior in a monkey group. *Primates* 9: 319–322.

Tolman, E. C.
1924   The inheritance of maze-learning ability in rats. *Journal of Comparative Psychology*, 4: 1–18.

Treisman, M.
1957a   Predation and the evolution of gregariousness. I. Models for concealment and evasion. *Animal Behaviour*, 23: 779–800.
1975b   Predation and the evolution of gregariousness. II. An economic model for predator-prey interaction, *Animal Behaviour*, 23: 801–825.

Trivers, R. L.
1971   The evolution of reciprocal altruism. *Quarterly Review of Biology*, 46: 35–57.
1972   Parental investment and sexual selection. In *Sexual Selection and the Descent of Man* (B. Campbell, ed.), pp. 136–179. Chicago: Aldine.
1974   Parent-offspring conflict. *American Zoologist*, 14: 249–264.

Trivers, R. L., and H. Hare
1976   Haplodiploidy and the evolution of the social insects. *Science*, 191: 249–263.

Trivers, R. L., and D. E. Willard
1973   Natural Selection of parental ability to vary the sex ratio of offspring. *Science*, 179: 90–92.

Turner, F. B., R. I. Jennrich, and J. D. Weintraub
1969   Home ranges and body size of lizards. *Ecology*, 50: 1076–1081.

Tuttle, R. H., ed.
1975   *Socioecology and Psychology of Primates*. Paris: Mouton.

Ulrich, R. E.
1966   Pain as a cause of aggression. *American Zoologist*, 6: 643–662.

Van Den Berghe, P.
1974   Bringing beasts back in: Toward a biosocial theory of aggression. *American Sociological Review*, 39: 778–788.
1975   *Man in Society*. New York: Elsevier.

Van Hooff, J. A. R. A. M.
1972   A comparative approach to the phylogeny of laughter and smiling. In *Non-Verbal Communication* (R. A. Hinde,

ed.), pp. 209–241. Cambridge: Cambridge University Press.

Van Lawick-Goodall, J.
1968   The behaviour of free-living chimpanzees in the Gombe Stream Reserve. *Animal Behaviour Monographs,* 1: 161–311.
1971   *In the Shadow of Man.* Boston: Houghton Mifflin.

Verner, J.
1964   Evolution of polygyny in the long-billed marsh wren. *Evolution,* 18: 252–261.

Waddington, C. H.
1960   Genetic assimilation. In *Advances in Genetics,* vol. 10 (E. W. Caspari and J. M. Thoday, eds.), pp. 257–293. New York: Academic.

Ward, P.
1965   Feeding ecology of the black-faced dioch *(Quelea quelea)* in Nigeria. *Ibis,* 107: 173–214.

Ward, P., and A. Zahavi
1973   The importance of certain assemblages of birds as "information-centres" for food finding. *Ibis,* 115: 517–534.

Watson, A., and D. Jenkins
1968   Experiments on population control by territorial behaviour in red grouse. *Journal of Animal Ecology,* 37: 595–614.

Watson, A., and G. R. Miller
1971   Territory size and aggression in a fluctuating red grouse population. *Journal of Animal Ecology,* 40: 367–383.

Watson, A., and R. Moss
1970   Dominance, spacing behaviour and aggression in relation to population limitation in vertebrates. In *Animal Populations in Relation to their Food Resources* (A. Watson, ed.), pp. 167–218. Oxford: Blackwell.

Watson, J. B.
1930   *Behaviorism.* New York: Norton.

Watts, C. R., and A. W. Stokes
1971   The social order of turkeys. *Scientific American,* 224: 112–118.

Weeden, J.S.
1965 Territorial behavior of the tree sparrow. *Condor,* 67: 193–209.

Weeden, J. S., and J. B. Falls
1959 Differential responses of male ovenbirds to recorded songs of neighboring and more distant individuals. *Auk,* 76: 343–351.

Weller, J. M.
1969 *The Course of Evolution.* New York: McGraw-Hill.

Welty, J.
1963 *The Life of Birds.* New York: Knopf.

West Eberhard, M. J.
1969 The social biology of polistine wasps. *Miscellaneous Publications, Museum of Zoology, University of Michigan, Ann Arbor,* 140: 1–101.
1975 The evolution of social behavior by kin selection. *Quarterly Review of Biology,* 50: 1–33.

Wickler, W.
1968 **Mimicry**. New York: McGraw-Hill.

Wiens, J. A.
1966 On group selection and Wynne-Edwards' hypothesis. *American Scientist,* 54: 273–287.

Wiley, R. H.
1973 Territoriality and non-random mating in sage grouse, *Centrocercus urophasianus. Animal Behaviour Monographs,* 6: 85–169.
1974 Evolution of social organization and life history patterns among grouse: *Tetraonidae. Quarterly Review of Biology,* 49: 201–227.

Williams, G. C.
1966a *Adaptation and Natural Selection.* Princeton N. J.: Princeton University Press.
1966b Natural selection, the costs of reproduction, and a refinement of Lack's principle. *American Naturalist,* 100: 687–690.
1971 *Group Selection.* Chicago: Aldine-Atherton.
1975 *Sex and Evolution.* Princeton, N. J.: Princeton University Press.

Wilson D. S.
1975   A theory of group selection. *Proceedings of the National Academy of Sciences,* 72: 143–146.

Wilson, E. O.
1968   The ergonomics of caste in the social insects. *American Naturalist,* 102: 41–66.
1971   *The Insect Societies.* Cambridge: Harvard University Press.
1973   Group selection and its significance for ecology. *BioScience,* 23: 631–638.
1975   *Sociobiology, The New Synthesis.* Cambridge: Harvard University Press.
1976   Academic vigilantism and the political significance of sociobiology. *BioScience,* 26: 183–190.

Wilson, E. O., and W. H. Bossert
1971   *A Primer of Population Biology.* Sunderland, Mass: Sinnauer Associates.

Wickler, W.
1968   *Mimicry in Plants and Animals.* New York: McGraw-Hill.

Wolf, L. L.
1975   "Prostitution" behavior in a tropical hummingbird. *Condor,* 77: 140–144.

Woolfenden, G. E.
1975   Florida scrub jay helpers at the nest. *Auk,* 92: 1–15.

Wright, S.
1969   *Evolution and the Genetics of Populations, Vol. 2: The Theory of Gene Frequencies.* Chicago: University of Chicago Press.

Wynne-Edwards, V. C.
1962   *Animal Dispersion in Relation to Social Behaviour.* New York: Hafner.

Yeaton, R. I., and M. L. Cody
1974   Competitive release in island song sparrow populations. *Theoretical Population Biology,* 5: 42–58.

Zahavi. A.
1975   Mate selection—a selection for a handicap. *Journal of Theoretical Biology,* 53: 205–214.

Zirkle, C.
1949   *Death of a Science in Russia.* Philadelphia: University of Pennsylvania Press.

Index

Ants, slave-making, 135
Archer, J., 218
Ardrey, R., 255
Armitage, K. B., 58, 170
Armstrong, E. A., 232
Armstrong, J. T., 253
Assortative mating, 31
Aunting, 109, 198

Baboons
    East African: flexibility in social
        structure of, 134; influence of
        predation on, 213; paternal
        behavior in, 192
    gelada, 114
    hamadryas, 198, 213
    home ranges of, 252
    oligarchies in, 238
Barash, D. P., 58, 62, 63, 65, 102,
    111, 115, 116, 122, 128, 139, 163,
    186, 218, 233, 251, 314
Barnett, S. A., 262
Bartholomew, G. A., 141
Bats: lekking behavior in, 166; roosts
    of, 118
Bears, aggression in, 225;
Behaviorism, 286
Bekoff, M., 129
Bentley, D. R., 46
Benzer, S., 45
Bernstein, I. S., 234
Biological conditioning of the envi-
    ronment, 121–2
Bipedalism, 24
Bison: aggression in, 216, 222;
    grouping and predation among,
    111–2
Blackbirds: inter-specific territorial-
    ity in, 263; role of courtship in,
    146; social systems of, 132–4; ter-
    ritory in, 274
Blake, J., 291
Bluebirds, mountain: absence of true
    altruism in, 93; response to adul-
    tery in, 63–6

Blurton-Jones, N., 277
Bobolinks, parental behavior in, 197
Bolles, R. C., 3
Boorman, S. A., 75–6
Bossert, W. H., 184
Bowerbirds, 153, 155
Bowlby, J., 305
Bradbury, J., 166
Brain size, evolution of, 16
Breeding, synchrony of, 119–20
Breland, K., 3, 36
Breland, M., 3, 36
Bronson, F. H., 262
Brothers, D. J., 84
Brown, J. L., 89, 91, 145, 199, 250,
    252, 254, 255
Brown, W. L., 263
Buechner, H. K., 166
Bumpus, H. C., 17–8
Buntings, indigo, 164–5
Burros, parent-offspring conflict in,
    207

Caciques, 135–6
Calhoun, J. B., 217
Campbell, D. T., 7, 103, 235, 313
Carey, M., 164
Caribou, 249
Carl, E. A., 109
Carlisle, R. T., 189
Carpenter, C. R., 72, 193, 255
Carrying capacity, 182
Caste ergonomics, 123
Caughley, G., 249
Cave-dwelling animals, 29
Cerebral cortex, 4
Chance, M. R. A., 231, 246
Character displacement, 263–5
Charnov, E. L., 68, 102
Cheating, relation to reciprocal al-
    truism, 95–6, 101, 109
Cheetahs, grouping in, 117
Chickadees, black-capped, grouping
    in, 116

Detour experiment, 3–5
De Vore, I., 192, 226, 313
Dilger, W., 47
Dispersal: correlating environment with age of, 59; density effects on, 217–8; r-selected species and, 184; strategies of, 109, 133, 248–9, 275
Dispersion, patterns of, 248–57
Distance species, 251
Distraction display, 186
Divorce, 294–95
DNA (Deoxyribosenucleic Acid), 47
Dobzhansky, T., 30, 34, 145, 288
Dogs
  African hunting: group hunting by, 119; group territories in, 269
  detour experiment, 3–5
Dollard, J. N., 221
Domestic fowl, selection for aggressiveness and mating behavior in, 45
Dominance: competition for, 23; relation of kin selection to, 241–2; site-dependent, 254; social tradition and, 124; strategies of, 209–46
Dominance by parents, consequences for success of offspring, 176–7
Downhower, J. F., 170
Driver, P. M., 112
Ducks: behavioral phylogenics of, 53; genetic bases for courtship in, 144
Dulosis, 135
Dunbar, E. P., 243
Dunbar, R. I. M., 243

Early experience, 131
Eaton, R., 117
Eaton, T. H., 34
Eberhard, W. G., 84
Ecology, evolutionary, 34
Economics, 317
Ehrenkrantz, J., 224
Ehrlich, P. R., 34

Ehrman, L., 18, 45, 47
Eibl-Eibesfeldt, I., 6, 227, 231, 260, 277, 320
Eiseley, L., 9
Eisenberg, J. F., 59, 105, 134, 217, 219, 225, 252
Elephants, 110
Elk, 249
Emigration, 31, 75
Emlen, J. M., 34, 180, 299
Emlen, S. T., 117–8
Environment, role in behavior, 39–52
Epideictic displays, 71–2
Erickson, C. J., 152
Erlenmeyer-Kimling, L., 45
Errington, P. L., 109
Estes, R. D., 133, 261
Ethograms, 38
Ethology, 6, 8; human, 277
Etkin, W., 145
Eusociality, 83–5
Evans, H. E., 53
Evolution: as a paradigm for behavior, 1–8; as a process, 9–34; biological vs cultural, 311–2, 318–24; compromise in, 50; definition of, 11; evidence for, 9; historical approach, 52–4; relevance to behavior, 35–69; synthetic theory of, 9, 30; techniques in studying behavior, 52–69
Ewer, R. F., 255
Extinction, 25, 27; group, 75–6

Fagen, R. M., 184
Falls, J. B., 257
Farner, D. S., 120
Finches, Galapagos, behavioral phylogenies of, 53
Fischer, G. J., 219
Fish: labrid, 178; stickleback, 190–1
Fitch, H. S., 252

Fitness: definition of, 32–3; inclusive, 63, 79–94; relative, 33
Fixed action pattern, 6, 277
Flickers, 148
Flies
    fruit: directional selection in, 15, 45; evolution of courtship in, 49–51; lekking behavior in, 166
    scorpion, courtship of, 54
Floaters, non-territorial, 271–5
Foraging, role of grouping in, 118–9
Fox, R., 7, 290
Foxes: communication in, 129; neighbor recognition by, 125, 128
Fraenkel, G. S., 2
Fretwell, S. D., 239, 274
Frogs, dendrobatid, 197
Fromm, E., 210
Fry, C. H., 91

Gadgil, M., 75, 184
Galusha, J., 146, 272
Gametes, difference between male and female, 147
Garcia, J., 3
Gardner, B. T., 130
Gardner, R. A., 130
Gartlan, J. S., 59, 133, 243
Gazelles
    Grant's: dominance in, 229; territories in, 260
    Thompson's, courtship in, 159
Geist, V., 26, 133, 142, 225, 231, 237, 238
Genetic assimilation, 48
Genetic drift, 31–2
Genetic relatedness, confidence of, 151–2, 189–94, 300–302
Genetic variability, importance of, 11–4, 19–21
Genotype, role in behavior, 39–52
Gibbons, 133, 192–4
Gilliard, E. T., 153

Giraffes, aggression among, 319–20
Glas, P., 272
Gorillas, 95, 199, 236; absence of territory in, 269
Goss-Custard, J. D., 118
Gouramis, kissing, 229
Grant, V., 34
Greenberg, B., 242
Greene, P. J., 314
Grooming, social, 236
Grouping: adaptive significance of, 104–36; breeding synchrony and, 119–21; classification of, 126–7; division of labor and, 122–3; evolution of in humans, 115; food-getting advantages of, 116–9; optimal size of, 266; role of predation in selecting for, 110–5; role of social tradition in, 125; significance for social tradition, 124–5
Groups: bachelor, 108, 275; clonal, 127; colonial, 127; mating, 126; survival, 126; unisexual, 127
Group selection. See Selection, group.
Grouse: black, age and reproductive success in, 240–1; sage, lekking behavior of, 167; sexual bimaturism in, 141
Guhl, A. M., 219, 237
Gulls: egg-shell removal by, 54–5; divorce among, 294; herring, 146, 272
Gunn, D. L., 2
Guthrie, E. R., 3

Hager, J. L., 6
Hailman, J. P., 120
Haldane, J. B. S., 30
Hall, E. T., 251
Hall, K. R. L., 192
Hamilton, W. D., 79, 84–5, 111, 175, 239, 310
Hamilton, W. J., 311

Mammals (*cont'd*)

offspring conflict in, 202–5; sexual dimorphism among, 113–5; sociality in, 105

Manning, A., 50

Marler, P. R., 251

Marmots: alarm calling by, 115; correlating behavior with environments, 57–62; paternal behavior in, 194–5

Martin, S. G., 55, 198

Mate selection: human strategies of, 289–93; male-female differences in, 147–8; strategies of, 144–60

Mather, K., 15

Mating strategies, individual differences in, 168–70

Mating systems: classification of, 161; influence of genetic relatedness upon, 171–2; strategies of, 160–72

Maynard Smith, J., 34, 79, 139, 171, 233

Mayr, E., 30, 34, 144, 265

McClearn, G. E., 49, 237

McGrew, W. C., 277

McKay, F. E., 242

McNab, B. K., 253

Mead, A. P., 246

Melanism, industrial, 16–7

Mendel, G., 30

Menopause, sociobiology of, 291

Meyerriecks, A. J., 253

Mice, selection for alcohol preference in, 45

Michael, R. P., 134

Michener, C. D., 84

Michener, G. R., 218

Migration, 249; proximate vs ultimate causation, 37

Milgram, S., 246

Mimicry, 34, 112; egg, 135–6

Minnows, 242

Mobbing, of predators, 113

Moehlman, P., 94, 207

Mohr, H., 113

Moment, G., 18

Monkeys

colobus, 243; howlers, 72; patas, 133; rhesus: crowding and aggression in, 216–7; isolation rearing of, 121; mother-infant separation in, 305; ritualization in, 129–30; urban vs rural, 134, 213

strangers and aggression in, 219

Monogamy, 160–3; evaluating adaptive significance of, 55–6, 105

Montagu, M. F. A., 210, 256

Moody, P. A., 9, 34

Morse, D. H., 116, 265

Moss, R., 273, 275

Muckenhirn, N. A., 134

Murray, B. G., 265

Murres, thick-billed, 214

Murton, R. K., 242

Mutation, 21–2, 24–5, 28, 31, 33, 45

Mutalism, 96–7

Myers, N., 106, 120

Nasrudin, 5–6

Native hens, Tasmanian, polyandry in, 90–1, 171

Neighbor recognition, 125, 128

Nepotism, biology of, 88–94, 309

Nest parasitism, 135–6, 151

Nice, M. M., 255

Nist, B., 215

Noble, G. K., 148

Nolan, V., 164

Operant conditioning, 3, 36

Optimal strategies, 68, 250; conflicts in, 170

Orang-utans: paternal behavior in, 196; oreopendolas, 135–6

Orians, G. H., 59, 132, 162, 163, 164, 250, 263

Orthogenesis, 24

Oryxes, Arabian, 240, 319–20
Otte, D., 130
Ovenbirds, territory and food supply of, 261
Oxen, musk, 311–2
Oystercatchers, learning in, 143

Paradigm, evolution as a, 1–8
Parental behavior: cooperation in feeding young, 55–6; costs and benefits of, 79–80; diversity in blackbirds, 132; human strategies of, 297–308; influence of presumed genetic relatedness on, 189–94; influence of reproductive potential on, 187–8; large male role in, 197; male-female differences in, 188–201, 289–93; offspring age and, 186–7; strategies of, 173–208
parental investment, 156–8, 160, 165, 166, 175, 180, 188–9, 202–6, 224, 290, 299–300, 303
Parental manipulation, 96
Parent-offspring conflict, 109, 201–8, 303–8
Parker, G. A., 147, 233
Parsons, P. A., 45, 47
Paternal behavior, 192–8
Paulson, D., 91
Pavlov, I., 2
Pearson, C. A., 252
Peek, F. W., 142, 274
Penguins: Adelie, 74; parental behavior in, 197
Petit, C., 18
Pheromones, 128, 259
Phylogenetic inertia, 125
Phylogeny, behavioral, 6, 52–3
Piaget, J., 288
Pianka, E. R., 34, 180
Pikas, aggression and dispersal in, 218
Pitelka, F. A., 59, 165, 254

Pliocene, 24
Plovers, golden, 249
Political science, 317
Polyandry, 90, 138, 157, 165
Polygyny, 138, 141–2, 157, 160–72, 196
Polygyny threshold, 163–5
Polymorphism, 20–1; behavioral, 171–2
Population genetics, 30–3
Population regulation, role of territoriality in, 271–5
Power, H., 64, 93
Prairie dogs: dispersal in, 218; territories of, 268
Predation, as a selecting agent, 110–5
Predators, 5; avoidance of, 38, 42; home ranges of, 253; mobbing of, 113; warning against, 109
Premack, D., 130
Price, G. R., 233
Primary socialization, 131
Primates, 59, 181; agonistic buffering in, 198–9; bachelor groups among, 275; dominance and roles in, 243; ecology of social systems in, 133–4; mating by subordinates in, 241; sexual dimorphism among, 113
Promiscuity, 160, 161, 165–9
Prosimians, 133
Prostitution behavior by hummingbirds, 159–60
Proximate causation, 37–8
Psychological castration, 243–4
Psychology, 1, 276, 317; comparative, 5–6; learning theorists in, 3; social, 6

r-selection. See Selection, r-.
Raccoons, neighbor recognition by, 125, 128
Ralph, C. J., 252
Rape, in mallards, 67–8
Reciprocity. See Altruism, reciprocal.
Recombination, sexual, 21–2, 28, 48

Reflex, 2
Reflexive fighting, 221
Reinforcement, 3, 36
Relationship, coefficient of, 85–7
Releasers, 65, 263, 277
Reproduction: capacity for, 10; optimum rate of, 72–3; sexual, pros and cons of, 138–40; strategies of, 137–72; stress of, 141; timing of, 140–3
Reproductive effort, strategies of, 179–88
Reproductive potential, influence upon parental investment, 186
Reproductive restraint, 70–3
Reproductive strategies, individual differences in, 168–70
Reproductive success, variance in, 156–7
Resources: territories as a function of, 132; competition for, 209–46, 247–75
Rhinoceros, black, territoriality in, 254; Indonesian, 106
Ricklefs, R. E., 34, 180
Ridpath, M. G., 90–1, 171
Ripley, S. D., 106, 150, 233
Ritualization, 129–30, 150, 237
Robbins, R., 80
Robertson, D. R., 178
Rodents, 105
Rogers, J., 234
Roth, H. D., 166
Rothenbuhler, W., 45
Rowell, T. E., 134, 198
Rowley, I., 92
Rudran, R., 134
Rumplestiltskin effect, 6

Sackett, G. P., 129
Sadleir, R. M. F. S., 218
Sahlins, M. D., 315–6
Sandpipers, Arctic, 59
Schaller, G. B., 101, 234, 236, 237
Schenkel, R., 254

Schoener, T. W., 68, 253
Scott, J. P., 221, 223
Sea lions, Stellar, 153
Seals: bachelor groups among, 275, elephant, 27, 142; fur, critical resources in, 213–4; sexual bimaturism in, 141
Searle, L. V., 49
Sebeok, T. H., 130
Selander, R. K., 123, 219, 270
Selection: artificial, 12–4, 45, 48–9; contrasted with Lamarckism, 27–30; creativity and, 24–5; definition of, 11–2; directional, 13–7, 23, 28, 45, 50, 263–4; disruptive, 18; effect on gene frequencies, 32–3; epigamic, 152–6, 163; for complex behaviors, 48; for detour ability, 4–5; frequency-dependent, 18; group, 74–9, 98–9, 101, 109, 123, 232, 248; group, for altruism in viruses, 76–7; individual vs group, 70–9; individual vs species benefit, 25–7; intra-sexual, 156–60, 163; K-, 182–5, 217; K-, and human behavior, 184–5; K-, and parental behavior, 199–201; K-, and territorial defense, 270; kin, 79–94, 98–9, 101, 109, 123, 198, 232, 241–2, 248, 281–2; mechanism of, 10–1; misconceptions concerning, 22–30; r-, 181–5, 217; response to, 43; sexual, 113, 152–60, 163, 165, 167, 289; stabilizing, 17–9, 51
Selection differential, 43
Selfishness, 70–103, 96–7, 101, 109, 152, 198, 232; social grouping and, 111
Seligman, M. E. P., 6
Selye, H., 245
Semelparity, 184
Separation, mother-infant, 305
Sex of offspring, strategies of choosing, 174–9

Sex ratio, 174–6, 270
Sexual bimaturism, 141–3, 270, 291
Sexual characteristics, secondary, significance of, 150
Sexual dimorphism: division of labor and, 123; role of predation in selecting for, 113–5
Sharpe, R. S., 46, 144
Sheep, bighorn, 26, 59, 142, 225, 231, 238
Shepher, J., 301
Sheppard, P., 21
Sherrington, E. S., 2
Shettleworth, S. J., 6
Shirer, S. W., 252
Siamangs, 192–4
Sickle-cell anemia, 20
Siegel, P., 45
Simpson, G. G., 9, 34
Singh, S. D., 134, 213
Skutch, A. F., 75, 91
Smith, A., 6
Smith, C. C., 117, 250
Smith, D., 90
Smith, N. G., 136
Smith, R., 190, 191
Snow, D. W., 72, 255
Social Darwinism, 7
Social facilitation, 121
Social systems: ecology of, 132–4; strategies of, 160–72
Social tradition, 124–5
Sociobiology: assumptions of, 33; central theorem of, 63, 277, 309; correlational approach to, 56–62, 132–4; definition of, 2; evaluative approach to, 54–6; historical approach to, 52–4; predictive approach to, 61–9, 276–324
Sociograms, 38
Sociology, 1, 6, 276, 317; relationship with evolution, 6–7
Socrates, 276
Southwick, C. H., 213, 216, 219
Sparr, E., 186

Sparrows
song: role of courtship in, 149; territories of, 255
white-crowned: parental behavior of, 187–8, 297–9; song learning by, 40
Speech, evolution of, 24
Spence, K. W., 3
Spencer-Booth, Y., 305
Spiess, E. B., 144
Spite, 101
Spurr, J., 82, 193, 239
Squirrels: Arctic ground, dispersal in, 109; detour experiment, 3–5; Richardson's ground, 218; solitary vs grouped living in, 117; territories of, 261
Stebbins, G. L., 34
Stefanski, R. A., 252
Stenger, J., 261
Stevenson-Hinde, J., 6
Stokes, A. W., 90, 171
Subordinate, compensations of being, 238–44
Sugiyama, Y., 100
Suzuki, A., 233
Swift, English, 72–3

Taber, R. D., 79
Tamarins, cotton-topped, paternal behavior of, 193
Taulman, J., 62, 195, 268
Taxes (tropisms), 2
Tenaza, R., 74, 121, 214
Territory: adaptive significance of, 258–9; competition for, 23; consequences of failure in defending, 269–75; consequences of success in defending, 269–70; defense of, reasons for, 257–69; definition of, 254; effects of crowding on, 271; group defense of, 265–7; in humans, 255–6, 267, 269; interspecific, 263; optimum size of, 261–2; population regulation and,